Water Supply Management and Urban Planning

Water Supply Management and Urban Planning

Edited by
George Dickens

Larsen & Keller
www.larsen-keller.com

Water Supply Management and Urban Planning
Edited by George Dickens
ISBN: 978-1-63549-723-6 (Hardback)

Larsen & Keller
Published by Larsen and Keller Education,
5 Penn Plaza,
19th Floor,
New York, NY 10001, USA

Cataloging-in-Publication Data

Water supply management and urban planning / edited by George Dickens.
p. cm.
Includes bibliographical references and index.
ISBN 978-1-63549-723-6
1. Water-supply--Management. 2. City planning. 3. Water resources development.
4. Municipal water supply. I. Dickens, George.
TD345 .W38 2018
333.91--dc23

The publisher's policy is to use permanent paper from mills that operate a sustainable forestry policy. Furthermore, the publisher ensures that the text paper and cover boards used have met acceptable environmental accreditation standards.

Printed and bound in China.

For more information regarding Larsen and Keller Education and its products, please visit the publisher's website www.larsen-keller.com

Table of Contents

Preface

Water supply management and urban planning are allied fields of study. Urban planning refers to the design, development, building, maintaining, planning the urban areas and cities. Water resource management is the distribution and planning of water resources in the most optimum way. This book is a compilation of chapters that discuss the most vital concepts in the field of urban planning with respect to water supply. It presents the complex subject in the most comprehensible and easy to understand language. The textbook is appropriate for those seeking detailed information in this area.

A detailed account of the significant topics covered in this book is provided below:

Chapter 1- Water supply is the distribution of water as a public utility by organizations or individuals. Other topics that have been discussed include water point mapping, non-revenue water, water supply network, etc. The chapter on water supply and sanitation offers an insightful focus, keeping in mind the complex subject matter.

Chapter 2- Aqueducts are canals, tunnels and other structures that are used to transport water. Water kiosk, spring supply, water tower, water tank, rainwater harvesting and water well are some of the water supply infrastructures. This chapter is an overview of the subject matter incorporating all the major aspects of water supply infrastructures.

Chapter 3- Sewage is a form of waste produced by people. It consists of greywater and blackwater. Sewage if untreatedt causes widespread water pollution. The process of removing pollutants from sewage water is known as sewage treatment. The chapter strategically encompasses and incorporates the treatment and management of sewage, providing a complete understanding.

Chapter 4- Water quality is the measurement of the condition of water. Water fluoridation, water purification, desalination, water treatment and reclaimed water are some of the topics discussed in the section. The aspects elucidated in this chapter are of vital importance, and provide a better understanding of water quality.

I would like to make a special mention of my publisher who considered me worthy of this opportunity and also supported me throughout the process. I would also like to thank the editing team at the back-end who extended their help whenever required.

Editor

1

An Introduction to Water Supply

Water supply is the distribution of water as a public utility by organizations or individuals. Other topics that have been discussed include water point mapping, non-revenue water, water supply network, etc. The chapter on water supply and sanitation offers an insightful focus, keeping in mind the complex subject matter.

Water Supply

Water from a tap - supplied by a pipe

Water supply is the provision of water by public utilities commercial organisations, community endeavors or by individuals, usually via a system of pumps and pipes. Irrigation is covered separately.

Global Access to Clean Water

In 2010, about 87% of the global population (5.9 billion people) had access to piped water supply through house connections or to an improved water source through other means than house, including standpipes, water kiosks, spring supplies and protected wells. However, about 13% (about 900 million people) did not have access to an improved water source and had to use unprotected wells or springs, canals, lakes or rivers for their water needs.

Shipot, an underground water source in Ukraine

A clean water supply - in particular water that is not polluted with fecal matter from lack of sanitation - is the single most important determinant of public health. Destruction of water supply and/or sanitation infrastructure after major catastrophes (earthquakes, floods, war, etc.) poses the immediate threat of severe epidemics of waterborne diseases, several of which can be life-threatening.

Technical Overview

Water supply systems get water from a variety of locations after appropriate treatment, including groundwater (aquifers), surface water (lakes and rivers), and the sea through desalination. The water treatment steps include, in most cases, purification, disinfection through chlorination and sometimes fluoridation. Treated water then either flows by gravity or is pumped to reservoirs, which can be elevated such as water towers or on the ground (for indicators related to the efficiency of drinking water distribution see non-revenue water). Once water is used, wastewater is typically discharged in a sewer system and treated in a sewage treatment plant before being discharged into a river, lake or the sea or reused for landscaping, irrigation or industrial use.

Use

In the U.S, the typical single family home uses about 138 gallons (521 litres) of water per day (2016 estimate) or 58.6 gallons per capita per day (222 liters). This includes several common residential end use purposes (in decreasing order) like toilet use, showers, tap (faucet) use, washing machine use, leaks, other (unidentified), baths, and dishwasher use.

Service Quality

Many of the 3.5 billion people having access to piped water receive a poor or very poor quality of service, especially in developing countries where about 80% of the world population lives. Water supply service quality has many dimensions: continuity; water quality; pressure; and the degree of responsiveness of service providers to customer complaints.

A young girl collects clean water from a communal water supply in Kawempe, Uganda.

Continuity of Supply

Continuity of water supply is taken for granted in most developed countries, but is a severe problem in many developing countries, where sometimes water is only provided for a few hours every day or a few days a week. It is estimated that about half of the population of developing countries receives water on an intermittent basis.

Water Quality

Drinking water quality has a micro-biological and a physico-chemical dimension. There are thousands of parameters of water quality. In public water supply systems water should, at a minimum, be disinfected—most commonly through the use of chlorination or the use of ultra violet light—or it may need to undergo treatment, especially in the case of surface water.

Water Pressure

Water pressures vary in different locations of a distribution system. Water mains below the street may operate at higher pressures, with a pressure reducer located at each point where the water enters a building or a house. In poorly managed systems, water pressure can be so low as to result only in a trickle of water or so high that it leads to damage to plumbing fixtures and waste of water. Pressure in an urban water system is typically maintained either by a pressurised water tank serving an urban area, by pumping the water up into a water tower and relying on gravity to maintain a constant pressure in the system or solely by pumps at the water treatment plant and repeater pumping stations.

1880s model of pumping engine, in Herne Bay Museum

Typical UK pressures are 4–5 bar (60-70 PSI) for an urban supply. However, some people can get over eight bars or below one bar. A single iron main pipe may cross a deep valley, it will have the same nominal pressure, however each consumer will get a bit more or less because of the hydrostatic pressure (about 1 bar/10 m height). So people at the bottom of a 100-foot (30 m) hill will get about 3 bars more than those at the top.

The effective pressure also varies because of the pressure loss due to supply resistance even for the same static pressure. An urban consumer may have 5 metres of 15 mm pipe running from the iron main, so the kitchen tap flow will be fairly unrestricted, so high flow. A rural consumer may have a kilometre of rusted and limed 22 mm iron pipe, so their kitchen tap flow will be small.

For this reason, the UK domestic water system has traditionally (prior to 1989) employed a "cistern feed" system, where the incoming supply is connected to the kitchen sink and also a header/storage tank in the attic. Water can dribble into this tank through a 12 mm pipe, plus ball valve, and then supply the house on 22 or 28 mm pipes. Gravity water has a small pressure (say 1/4 bar in the bathroom) so needs wide pipes allow higher flows. This is fine for baths and toilets but is frequently inadequate for showers. A booster pump or a hydrophore is installed to increase and maintain pressure. For this reason urban houses are increasingly using mains pressure boilers (combies) which take a long time to fill a bath but suit the high back pressure of a shower.

Institutional Responsibility and Governance

A great variety of institutions have responsibilities in water supply. A basic distinction is between institutions responsible for policy and regulation on the one hand; and institutions in charge of providing services on the other hand.

Policy and Regulation

Water supply policies and regulation are usually defined by one or several Ministries, in consultation with the legislative branch. In the United States the United

States Environmental Protection Agency, whose administrator reports directly to the President, is responsible for water and sanitation policy and standard setting within the executive branch. In other countries responsibility for sector policy is entrusted to a Ministry of Environment (such as in Mexico and Colombia), to a Ministry of Health (such as in Panama, Honduras and Uruguay), a Ministry of Public Works (such as in Ecuador and Haiti), a Ministry of Economy (such as in German states) or a Ministry of Energy (such as in Iran). A few countries, such as Jordan and Bolivia, even have a Ministry of Water. Often several Ministries share responsibilities for water supply. In the European Union, important policy functions have been entrusted to the supranational level. Policy and regulatory functions include the setting of tariff rules and the approval of tariff increases; setting, monitoring and enforcing norms for quality of service and environmental protection; benchmarking the performance of service providers; and reforms in the structure of institutions responsible for service provision. The distinction between policy functions and regulatory functions is not always clear-cut. In some countries they are both entrusted to Ministries, but in others regulatory functions are entrusted to agencies that are separate from Ministries.

Regulatory Agencies

Dozens of countries around the world have established regulatory agencies for infrastructure services, including often water supply and sanitation, in order to better protect consumers and to improve efficiency. Regulatory agencies can be entrusted with a variety of responsibilities, including in particular the approval of tariff increases and the management of sector information systems, including benchmarking systems. Sometimes they also have a mandate to settle complaints by consumers that have not been dealt with satisfactorily by service providers. These specialized entities are expected to be more competent and objective in regulating service providers than departments of government Ministries. Regulatory agencies are supposed to be autonomous from the executive branch of government, but in many countries have often not been able to exercise a great degree of autonomy. In the United States regulatory agencies for utilities have existed for almost a century at the level of states, and in Canada at the level of provinces. In both countries they cover several infrastructure sectors. In many U.S. states they are called Public Utility Commissions. For England and Wales, a regulatory agency for water (OFWAT) was created as part of the privatization of the water industry in 1989. In many developing countries, water regulatory agencies were created during the 1990s in parallel with efforts at increasing private sector participation. (for example, Water and sanitation in Latin America and the regional association of water regulatory agencies ADERASA).

Many countries do not have regulatory agencies for water. In these countries service providers are regulated directly by local government, or the national government. This is, for example, the case in the countries of continental Europe, in China and India.

Service Provision

Water supply service providers, which are often utilities, differ from each other in terms of their geographical coverage relative to administrative boundaries; their sectoral coverage; their ownership structure; and their governance arrangements.

Geographical Coverage

The sole water supply of this section of Wilder, Tennessee, 1942.

Many water utilities provide services in a single city, town or municipality. However, in many countries municipalities have associated in regional or inter-municipal or multi-jurisdictional utilities to benefit from economies of scale. In the United States these can take the form of special-purpose districts which may have independent taxing authority. An example of a multi-jurisdictional water utility in the United States is WASA, a utility serving Washington, DC and various localities in the state of Maryland. Multi-jurisdictional utilities are also common in Germany, where they are known as "Zweckverbaende", in France and in Italy.

In some federal countries, there are water service providers covering most or all cities and towns in an entire state, such as in all states of Brazil and some states in Mexico. In England and Wales, water supply and sewerage is supplied almost entirely through ten regional companies. Some smaller countries, especially developed countries, have established service providers that cover the entire country or at least most of its cities and major towns. Such national service providers are especially prevalent in West Africa and Central America, but also exist, for example, in Tunisia, Jordan and Uruguay. In rural areas, where about half the world population lives, water services are often not provided by utilities, but by community-based organizations which usually cover one or sometimes several villages.

Sector Coverage

Some water utilities provide only water supply services, while sewerage is under the responsibility of a different entity. This is for example the case in Tunisia. However, in most cases water utilities also provide sewer and sewage treatment services. In some cities or countries utilities also distribute electricity. In a few cases such multi-utilities also collect solid waste and provide local telephone services. An example of such an integrated utility can be found in the Colombian city of Medellín. Utilities that provide water, sanitation and electricity can be found in Frankfurt, Germany (Mainova), in Casablanca, Morocco and in Gabon in West Africa. Multi-utilities provide certain benefits such as common billing and the option to cross-subsidize water services with revenues from electricity sales, if permitted by law.

Ownership and Governance Arrangements

Water supply providers can be either public, private, mixed or cooperative. Most urban water supply services around the world are provided by public entities. As Willem-Alexander, Prince of Orange (2002) stated, "The water crisis that is affecting so many people is mainly a crisis of governance — not of water scarcity." The introduction of cost-reflective tariffs together with cross-subsidization between richer and poorer consumers is an essential governance reform in order to reduce the high levels of Unaccounted-for Water (UAW) and to provide the finance needed to extend the network to those poorest households who remain unconnected. Partnership arrangements between the public and private sector can play an important role in order to achieve this objective.

Private Sector Participation

An estimated 10 percent of urban water supply is provided by private or mixed public-private companies, usually under concessions, leases or management contracts. Under these arrangements the public entity that is legally responsible for service provision delegates certain or all aspects of service provision to the private service provider for a period typically ranging from 4 to 30 years. The public entity continues to own the assets. These arrangements are common in France and in Spain. Only in few parts of the world water supply systems have been completely sold to the private sector (privatization), such as in England and Wales as well as in Chile. The largest private water companies in the world are Suez and Veolia Environnement from France; *Aguas de Barcelona* from Spain; and Thames Water from the UK, all of which are engaged internationally. In recent years, a number of cities have reverted to the public sector in a process called "remunicipalization".

Public Water Service Provision

90% of urban water supply and sanitation services are currently in the public sector. They are owned by the state or local authorities, or also by collectives or cooperatives. They run without an aim for profit but are based on the ethos of providing a common good considered to be of public interest. In most middle and low-income countries, these publicly owned and managed water providers can be inefficient as a result of political interference, leading to over-staffing and low labor productivity. Ironically, the main losers from this institutional arrangement are the urban poor in these countries. Because they are not connected to the network, they end up paying far more per liter of water than do more well-off households connected to the network who benefit from the implicit subsidies that they receive from loss-making utilities. The fact that we are still so far from achieving universal access to clean water and sanitation shows that public water authorities, in their current state, are not working well enough. Yet some are being very successful and are modelling the best forms of public management. As Ryutaro Hashimoto, former Japanese Prime Minister, notes: "Public water services currently provide more than 90 percent of water supply in the world. Modest improvement in public water operators will have immense impact on global provision of services."

Governance Arrangements

Governance arrangements for both public and private utilities can take many forms (Kurian and McCarney, 2010). Governance arrangements define the relationship between the service provider, its owners, its customers and regulatory entities. They determine the financial autonomy of the service provider and thus its ability to maintain its assets, expand services, attract and retain qualified staff, and ultimately to provide high-quality services. Key aspects of governance arrangements are the extent to which the entity in charge of providing services is insulated from arbitrary political intervention; and whether there is an explicit mandate and political will to allow the service provider to recover all or at least most of its costs through tariffs and retain these revenues. If water supply is the responsibility of a department that is integrated in the administration of a city, town or municipality, there is a risk that tariff revenues are diverted for other purposes. In some cases, there is also a risk that staff are appointed mainly on political grounds rather than based on their professional credentials.

Standardization

International standards for water supply system are covered by International Classification of Standards (ICS) 91.140.60.

Comparing the Performance of Water and Sanitation Service Providers

Comparing the performance of water and sanitation service providers (utilities) is needed, because the sector offers limited scope for direct competition (natural monopoly). Firms operating in competitive markets are under constant pressure to out perform each other. Water utilities are often sheltered from this pressure, and it frequently shows: some utilities are on a sustained improvement track, but many others keep falling further behind best practice. Benchmarking the performance of utilities allows to simulate competition, establish realistic targets for improvement and create pressure to catch up with better utilities. Information on benchmarks for water and sanitation utilities is provided by the International Benchmarking Network for Water and Sanitation Utilities.

Financial Aspects

Costs and Financing

The cost of supplying water consists, to a very large extent, of fixed costs (capital costs and personnel costs) and only to a small extent of variable costs that depend on the amount of water consumed (mainly energy and chemicals). The full cost of supplying water in urban areas in developed countries is about US$1–2 per cubic meter depending on local costs and local water consumption levels. The cost of sanitation (sewerage and wastewater treatment) is another US$1–2 per cubic meter. These costs are somewhat lower in developing countries. Throughout the world, only part of these costs is usually billed to consumers, the remainder being financed through direct or indirect subsidies from local, regional or national governments.

Besides subsidies water supply investments are financed through internally generated revenues as well as through debt. Debt financing can take the form of credits from commercial Banks, credits from international financial institutions such as the World Bank and regional development banks (in the case of developing countries), and bonds (in the case of some developed countries and some upper middle-income countries).

Tariffs

Almost all service providers in the world charge tariffs to recover part of their costs. According to estimates by the World Bank the average (mean) global water tariff is US$0.53 per cubic meter. In developed countries the average tariff is US$1.04, while it is only U$0.11 in the poorest developing countries. The lowest tariffs in developing countries are found in South Asia (mean of US$0.09/m3), while the highest are found in Latin America (US$0.41/m3). Data for 132 cities were assessed. The tariff is estimate for a consumption level of 15 cubic meters per month. Few utilities do recover all their costs. According to the same World Bank study only 30% of utilities globally, and only 50% of utilities in developed countries, generate sufficient revenue to cover operation, maintenance and partial capital costs.

According to another study undertaken in 2006 by NUS Consulting, the average water and sewerage tariff in 14 mainly OECD countries excluding VAT varied between US$0.66 per cubic meter in the United States and the equivalent of US$2.25 per cubic meter in Denmark. However, water consumption is much higher in the US than in Europe. Therefore, residential water bills may be very similar, even if the tariff per unit of consumption tends to be higher in Europe than in the US.

A typical family on the US East Coast paid between US$30 and US$70 per month for water and sewer services in 2005.

In developing countries, tariffs are usually much further from covering costs. Residential water bills for a typical consumption of 15 cubic meters per month vary between less than US$1 and US$12 per month.

Water and sanitation tariffs, which are almost always billed together, can take many different forms. Where meters are installed, tariffs are typically volumetric (per usage), sometimes combined with a small monthly fixed charge. In the absence of meters, flat or fixed rates — which are independent of actual consumption — are being charged. In developed countries, tariffs are usually the same for different categories of users and for different levels of consumption.

In developing countries, the situation is often characterized by cross-subsidies with the intent to make water more affordable for residential low-volume users that are assumed to be poor. For example, industrial and commercial users are often charged higher tariffs than public or residential users. Also, metered users are often charged higher tariffs for higher levels of consumption (increasing-block tariffs). However, cross-subsidies between residential users do not always reach their objective. Given the overall low level of water tariffs in developing countries even at higher levels of consumption, most consumption subsidies benefit the wealthier segments of society. Also, high industrial and commercial tariffs can provide an incentive for these users to supply water from other sources than the utility (own wells, water tankers) and thus actually erode the utility's revenue base.

Metering

Metering of water supply is usually motivated by one or several of four objectives: First, it provides an incentive to conserve water which protects water resources (environmental objective). Second, it can postpone costly system expansion and saves energy and chemical costs (economic objective). Third, it allows a utility to better locate distribution losses (technical objective). Fourth, it allows to charge for water based on use, which is perceived by many as the fairest way to allocate the costs of water supply to users. Metering is considered good practice in water supply and is widespread in developed countries, except for the United Kingdom. In developing countries it is estimated that half of all urban water supply systems are metered and the tendency is increasing.

Water meters are read by one of several methods:

- the water customer writes down the meter reading and mails in a postcard with this info to the water department;
- the water customer writes down the meter reading and uses a phone dial-in system to transfer this info to the water department;

- the water customer logs into the website of the water supply company, enters the address, meter ID and meter readings;
- a meter reader comes to the premises and enters the meter reading into a handheld computer;
- the meter reading is echoed on a display unit mounted to the outside of the premises, where a meter reader records them;
- a small radio is hooked up to the meter to automatically transmit readings to corresponding receivers in handheld computers, utility vehicles or distributed collectors;
- a small computer is hooked up to the meter that can either dial out or receive automated phone calls that give the reading to a central computer system.

A typical residential water meter

Most cities are increasingly installing Automatic Meter Reading (AMR) systems to prevent fraud, to lower ever-increasing labor and liability costs and to improve customer service and satisfaction.

Health Aspects

Outbreaks of Diseases Due to Contaminated Water Supply

Water supply can get contaminated by pathogens which may originate from human excreta, for example due to a break-down or design fault in the sanitation system, or by chemical contaminants.

Examples of contamination include:

- In 1854, a cholera outbreak in London's Soho district was identified by Dr. John Snow as originating from contaminated water from the Broad street pump. This can be regarded as a founding event of the science of epidemiology.

- In 1980, a hepatitis A surge due to the consumption of water from a feces-contaminated well, in Pennsylvania
- In 1987, a cryptosporidiosis outbreak is caused by the public water supply of which the filtration was contaminated, in western Georgia
- In 1993, Milwaukee Cryptosporidium outbreak
- An outbreak of typhoid fever in northern Israel, which was associated with the contaminated municipal water supply
- In 1997, 369 cases of cryptosporidiosis occurred, caused by a contaminated fountain in the Minnesota zoo. Most of the sufferers were children
- In 1998, a non-chlorinated municipal water supply was blamed for a campylobacteriosis outbreak in northern Finland
- In 2000, a gastroenteritis outbreak that was brought by a non-chlorinated community water supply, in southern Finland
- In 2000, an E. coli outbreak occurred in Walkerton, Ontario, Canada. Seven people died from drinking contaminated water. Hundreds suffered from the symptoms of the disease, not knowing if they too would die.
- In 2004, contamination of the community water supply, serving the Bergen city centre of Norway, was later reported after the outbreak of waterborne giardiasis
- In 2007, contaminated drinking water was pinpointed which had led to the outbreak of gastroenteritis with multiple aetiologies in Denmark

Examples of chemical contamination include:

- In 1988, many people were poisoned in Camelford, when a worker put 20 tonnes of aluminium sulphate in the wrong tank.
- In 1993, a fluoride poisoning outbreak resulting from overfeeding of fluoride, in Mississippi

History

Throughout history, people have devised systems to make getting and using water more convenient. Early Rome had indoor plumbing, meaning a system of aqueducts and pipes that terminated in homes and at public wells and fountains for people to use.

Until the Enlightenment era, little progress was made in water supply and sanitation and the engineering skills of the Romans were largely neglected throughout Europe. It was in the 18th century that a rapidly growing population fueled a boom in the establishment of private water supply networks in London. The Chelsea Waterworks Company was established in 1723 "for the better supplying the City and Liberties of Westminster and parts adjacent with water". Other waterworks were established in London, including at West Ham in 1743, at Lea Bridge before 1767, Lambeth Waterworks Company in 1785, West Middlesex Waterworks Company in 1806 and Grand Junction Waterworks Company in 1811.

The S-bend pipe was invented by Alexander Cummings in 1775 but became known as the U-bend following the introduction of the U-shaped trap by Thomas Crapper in 1880. The first screw-down water tap was patented in 1845 by Guest and Chrimes, a brass foundry in Rotherham.

Wasserkunst and fountain from 1602 in Wismar, Germany.
It's an example of pre-industrialization waterworks and fountain.

Chelsea Waterworks, 1752. Two Newcomen beam engines pumped
Thames water from a canal to reservoirs at Green Park and Hyde Park.

In ancient Peru, the Nazca people employed a system of interconnected wells and an underground watercourse known as puquios. In Spain and Spanish America, a community operated watercourse known as an acequia, combined with a simple sand filtration system, provided potable water. Beginning in the Roman era a water wheel device known as a noria supplied water to aqueducts and other water distribution systems in major cities in Europe and the Middle East. London water supply infrastructure developed over many centuries from early mediaeval conduits, through major 19th century treatment works built in response to cholera threats, to modern large scale reservoirs.

Water towers appeared around the late 19th century, as building height rose, and steam, electric and diesel-powered water pumps became available. As skyscrapers appeared, they needed rooftop water towers.

The technique of purification of drinking water by use of compressed liquefied chlorine gas was developed in 1910 by U.S. Army Major (later Brig. Gen.) Carl Rogers Darnall (1867–1941), Professor of Chemistry at the Army Medical School. Shortly thereafter, Major (later Col.) William J. L. Lyster (1869–1947) of the Army Medical Department used a solution of calcium hypochlorite in a linen bag to treat water. For many decades, Lyster's method remained the standard for U.S. ground forces in the field and in camps, implemented in the form of the familiar Lyster Bag (also spelled Lister Bag). Darnall's work became the basis for present day systems of municipal water '*purification*'.

Desalination appeared during the late 20th century, and is still limited to a few areas.

During the beginning of the 21st Century, especially in areas of urban and suburban population centers, traditional centralized infrastructure have not been able to supply sufficient quantities of water to keep up with growing demand. Among several options that have been managed are the extensive use of desalination technology, this is especially prevalent in coastal areas and in "dry" countries like Australia. Decentralization of water infrastructure has grown extensively as a viable solution including Rainwater harvesting and Stormwater harvesting where policies are eventually tending towards a more rational use and sourcing of water incorporation concepts such as "Fit for Purpose". Emirians have the highest per capita water consumption rate in the world, at 133 gallons.

Water Treatment

Original map by John Snow showing the clusters of cholera cases in the London epidemic of 1854.

The first documented use of sand filters to purify the water supply dates to 1804, when the owner of a bleachery in Paisley, Scotland, John Gibb, installed an experimental filter, selling his unwanted surplus to the public. The first treated public water supply in the world was installed by engineer James Simpson for the Chelsea Waterworks Company in London in 1829. The practice of water treatment soon became mainstream, and the virtues of the system were made starkly apparent after the investigations of the physician John Snow during the 1854 Broad Street cholera outbreak demonstrated the role of the water supply in spreading the cholera epidemic.

The Metropolis Water Act introduced regulation of the water supply companies in London, including minimum standards of water quality for the first time. The Act "made provision for securing the supply to the Metropolis of pure and wholesome water", and required that all water be "effectually filtered" from 31 December 1855. This legislation set a worldwide precedent for similar state public health interventions across Europe.

Permanent water chlorination began in 1905, when a faulty slow sand filter and a contaminated water supply led to a serious typhoid fever epidemic in Lincoln, England. Dr. Alexander Cruickshank Houston used chlorination of the water to stem the epidemic. His installation fed a concentrated solution of chloride of lime to the water being treated. The first continuous use of chlorine in the United States for disinfection took place in 1908 at Boonton Reservoir (on the Rockaway River), which served as the supply for Jersey City, New Jersey. Desalination appeared during the late 20th century, and is still limited to a few areas.

The technique of purification of drinking water by use of compressed liquefied chlorine gas was developed by a British officer in the Indian Medical Service, Vincent B. Nesfield, in 1903. U.S. Army Major Carl Rogers Darnall, Professor of Chemistry at the Army Medical School, gave the first practical demonstration of this in 1910. This work became the basis for present day systems of municipal water *purification*.

Society and Culture

Women and Water Supply Issues in Developing Countries

Water supply issues have specific adverse effects on women in developing nations. Women are often the primary family member responsible for providing water as well as collecting it. Inclusion of women in the design and implementation of water supply projects is an area of concern currently being addressed by multiple world organizations.

Lawsuits

- Arizona v. California
- Colorado River Water Conservation District v. United States
- Kansas v. Colorado
- Tahoe-Sierra Preservation Council, Inc. v. Tahoe Regional Planning Agency
- Wisconsin v. Illinois
- Wyoming v. Colorado

Water Supply Network

A water supply system or water supply network is a system of engineered hydrologic and hydraulic components which provide water supply. A water supply system typically includes:

1. A drainage basin.
2. A raw water collection point (above or below ground) where the water accumulates, such as a lake, a river, or groundwater from an underground aquifer. Raw water may be transferred using uncovered ground-level aqueducts, covered tunnels or underground water pipes to water purification facilities.
3. Water purification facilities. Treated water is transferred using water pipes (usually underground).
4. Water storage facilities such as reservoirs, water tanks, or water towers. Smaller water systems may store the water in cisterns or pressure vessels. Tall buildings may also need to store water locally in pressure vessels in order for the water to reach the upper floors.
5. Additional water pressurizing components such as pumping stations may need to be situated at the outlet of underground or above ground reservoirs or cisterns (if gravity flow is impractical).
6. A pipe network for distribution of water to the consumers (which may be private houses or industrial, commercial or institution establishments) and other usage points (such as fire hydrants).
7. Connections to the sewers (underground pipes, or aboveground ditches in some developing countries) are generally found downstream of the water consumers, but the sewer system is considered to be a separate system, rather than part of the water supply system.

Water Abstraction and Raw Water Transfer

Raw water (untreated) is collected from a surface water source (such as an intake on a lake or a river) or from a groundwater source (such as a water well drawing from an underground aquifer) within the watershed that provides the water resource.

The raw water is transferred to the water purification facilities using uncovered aqueducts, covered tunnels or underground water pipes.

Water Treatment

Virtually all large systems must treat the water; a fact that is tightly regulated by global, state and federal agencies, such as the World Health Organization (WHO) or the United States Environmental Protection Agency (EPA). Water treatment must occur before the product reaches the consumer and afterwards (when it is discharged again). Water purification usually occurs close to the final delivery points to reduce pumping costs and the chances of the water becoming contaminated after treatment.

Traditional surface water treatment plants generally consists of three steps: clarification, filtration and disinfection. Clarification refers to the separation of particles (dirt, organic matter, etc.) from the water stream. Chemical addition (i.e. alum, ferric chloride) destabilizes the particle charges and prepares them for clarification either by settling or floating out of the water stream. Sand, anthracite or activated carbon filters refine the water stream, removing smaller particulate matter. While other methods of disinfection exist, the preferred method is via chlorine addition. Chlorine effectively kills bacteria and most viruses and maintains a residual to protect the water supply through the supply network.

Water Distribution Network

The product, delivered to the point of consumption, is called potable water if it meets the water quality standards required for human consumption.

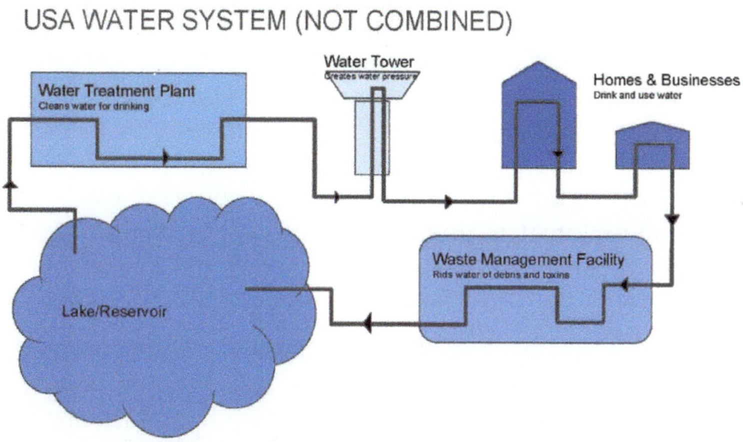

USA Not Combined City Water System

The water in the supply network is maintained at positive pressure to ensure that water reaches all parts of the network, that a sufficient flow is available at every take-off point and to ensure that untreated water in the ground cannot enter the network. The water is typically pressurised by pumps that pump water into storage tanks constructed at the highest local point in the network. One network may have several such service reservoirs.

In small domestic systems, the water may be pressurised by a pressure vessel or even by an underground cistern (the latter however does need additional pressurizing). This eliminates the need of a water-tower or any other heightened water reserve to supply the water pressure.

Pressurizing the water is required between the small water reserve and the end-user.

These systems are usually owned and maintained by local governments, such as cities, or other public entities, but are occasionally operated by a commercial enterprise. Water supply networks are part of the master planning of communities, counties, and municipalities. Their planning and design requires the expertise of city planners and civil engineers, who must consider many factors, such as location, current demand, future growth, leakage, pressure, pipe size, pressure loss, fire fighting flows, etc. — using pipe network analysis and other tools.

As water passes through the distribution system, the water quality can degrade by chemical reactions and biological processes. Corrosion of metal pipe materials in the distribution system can cause the release of metals into the water with undesirable aesthetic and health effects. Release of iron from unlined iron pipes can result in customer reports of "red water" at the tap . Release of copper from copper pipes can result in customer reports of "blue water" and/or a metallic taste. Release of lead can occur from the solder used to join copper pipe together or from brass fixtures. Copper and lead levels at the consumer's tap are regulated to protect consumer health.

Utilities will often adjust the chemistry of the water before distribution to minimize its corrosiveness. The simplest adjustment involves control of pH and alkalinity to produce a water that tends to passivate corrosion by depositing a layer of calcium carbonate. Corrosion inhibitors are often added to reduce release of metals into the water. Common corrosion inhibitors added to the water are phosphates and silicates.

Maintenance of a biologically safe drinking water is another goal in water distribution. Typically, a chlorine based disinfectant, such as sodium hypochlorite or monochloramine is added to the water as it leaves the treatment plant. Booster stations can be placed within the distribution system to ensure that all areas of the distribution system have adequate sustained levels of disinfection.

Topologies of Water Distribution Networks

Like electric power lines, roads, and microwave radio networks, water systems may have a loop or branch network topology, or a combination of both. The piping networks are circular or rectangular. If any one section of water distribution main fails or needs repair, that section can be isolated without disrupting all users on the network.

Most systems are divided into zones. Factors determining the extent or size of a zone can include hydraulics, telemetry systems, history, and population density. Sometimes systems are designed for a specific area then are modified to accommodate development. Terrain affects hydraulics and some forms of telemetry. While each zone may operate as a stand-alone system, there is usually some arrangement to interconnect zones in order to manage equipment failures or system failures.

Water Network Maintenance

Water supply networks usually represent the majority of assets of a water utility. Systematic documentation of maintenance works using a computerized maintenance management system (CMMS) is a key to a successful operation of a water utility.

Clean drinking water is essential to human life.

Sustainable Urban Water Supply

A sustainable urban water supply network covers all the activities related to provision of potable water. Sustainable development is of increasing importance for the water supply to urban areas. Incorporating innovative water technologies into water supply systems improves water supply from sustainable perspectives. The development of innovative water technologies provides flexibility to the water supply system, generating a fundamental and effective means of sustainability based on an integrated real options approach.

Water is an essential natural resource for human existence. It is needed in every industrial and natural process, for example, it is used for oil refining, for liquid-liquid extraction in hydro-metallurgical processes, for cooling, for scrubbing in the iron and the steel industry and for several operations in food processing facilities , etc. It is necessary to adopt a new approach to design urban water supply networks; water shortages are expected in the forthcoming decades and environmental regulations for water utilization and waste-water disposal are increasingly stringent.

To achieve a sustainable water supply network, new sources of water are needed to be developed, and to reduce environmental pollution.

The price of water is increasing, so less water must be wasted and actions must be taken to prevent pipeline leakage. Shutting down the supply service to fix leaks is less and less tolerated by consumers. A sustainable water supply network must monitor the freshwater consumption rate and the waste-water generation rate.

Many of the urban water supply networks in developing countries face problems related to population increase, water scarcity, and environmental pollution.

Population Growth

In the year 1900 just 13% of the global population lived in cities. This percentage has been rising, and in 2005 49% of the global population lived in urban areas. In 2030 it is predicted, that this statistic will rise to 60% . Attempts to expand water supply by governments are costly and often not sufficient. The building of new illegal settlements makes it hard to map, and make connections to, the water supply, and leads to inadequate water management . In 2002, there were 158 million people with inadequate water supply. An increasing number of people live in slums, in inadequate sanitary conditions, and are therefore at risk of disease.

Water Scarcity

Potable water is not well distributed in the world. 1.8 million deaths are attributed to unsafe water supplies every year, according to the WHO . Many people do not have any access, or do not have access to quality and quantity of potable water, though water itself is abundant. Poor people in developing countries can be close to major rivers, or be in high rainfall areas, yet not have access to potable water at all. There are also people living where lack of water creates millions of deaths every year.

Where the water supply system cannot reach the slums, people manage to use hand pumps, to reach the pit wells, rivers, canals, swamps and any other source of water. In most cases the water quality is unfit for human consumption. The principal cause of water scarcity is the growth in demand. Water is taken from remote areas to satisfy the needs of urban areas. Another reason for water scarcity is climate change: precipitation patterns have changed; rivers have decreased their flow; lakes are drying up; and aquifers are being emptied.

Governmental Issues

In developing countries many governments are corrupt and poor and they respond to these problems with frequently changing policies and non clear agreements. Water demand exceeds supply, and household and industrial water supplies are prioritised over other uses, which leads to water stress. Potable water has a price in the market; water often becomes a business for private companies, which earn a profit by putting a higher price on water, which imposes a barrier for lower-income people. The Millennium Development Goals propose the changes required.

In advanced economies, the problems are about optimising existing supply networks. These economies have usually had continuing evolution, which allowed them to construct infrastructure to supply water to people. The European Union has developed a set of rules and policies to overcome expected future problems.

There are many international documents with interesting, but not very specific, ideas and therefore they are not put into practice . Recommendations have been made by the United Nations, such as the Dublin Statement on Water and Sustainable Development.

Optimizing the Water Supply Network

The yield of a system can be measured by either its value or its net benefit. For a water supply system, the true value or the net benefit is a reliable water supply service having adequate quantity and good quality of the product. For example, if the existing water supply of a city needs to be extended to supply a new municipality, the impact of the new branch of the system must be designed to supply the new needs, while maintaining supply to the old system.

Single-objective Optimization

The design of a system is governed by multiple criteria, one being cost. If the benefit is *fixed*, the least cost design results in maximum benefit. However, the least cost approach normally results in a *minimum capacity* for a water supply network. A minimum cost model usually searches for the least cost solution (in pipe sizes), while satisfying the hydraulic constraints such as: required output pressures, maximum pipe flow rate and pipe flow velocities. The cost is a function of pipe diameters; therefore the optimization problem consists of finding a minimum cost solution by optimising pipe sizes to provide the minimum acceptable capacity.

Multi-objective Optimization

However, according to the authors of the paper entitled, "Method for optimizing design and rehabilitation of water distribution systems", "the least capacity is not a desirable solution to a sustainable water supply network in a long term, due to the uncertainty of the future demand" . It is preferable to provide extra pipe capacity to cope with unexpected demand growth and with water outages. The problem changes from a single objective optimization problem (minimizing cost), to a multi-objective optimization problem (minimizing cost and maximizing flow capacity).

Weighted Sum Method

To solve a multi-objective optimization problem, it is necessary to convert the problem into a single objective optimization problem, by using adjustments, such as a weighted sum of objectives, or an ε-constraint method. The weighted sum approach gives a certain weight to the different objectives, and then factors in all these weights to form a single objective function that can be solved by single factor optimization. This method is not entirely satisfactory, because the weights cannot be correctly chosen, so this approach cannot find the optimal solution for all the original objectives.

The Constraint Method

The second approach (the constraint method), chooses one of the objective functions as the single objective, and the other objective functions are treated as constraints with a limited value. However, the optimal solution depends on the pre-defined constraint limits.

Sensitivity Analysis

The multiple objective optimization problems involve computing the tradeoff between the costs and benefits resulting in a set of solutions that can be used for sensitivity analysis and tested in different scenarios. But there is no single optimal solution that will satisfy the global optimality of both objectives. As both objectives are to some extent contradictory, it is not possible to improve one objective without sacrificing the other. It is necessary in some cases use a different approach,(e.g. Pareto Analysis), and choose the best combination.

Operational Constraints

Returning to the cost objective function, it cannot violate any of the operational constraints. Generally this cost is dominated by the energy cost for pumping. "The operational constraints include the standards of customer service, such us: the minimum delivered pressure, in addition to the physical constraints such us the maximum and the minimum water levels in storage tanks to prevent overtopping and emptying respectively."

In order to optimize the operational performance of the water supply network, at the same time as minimizing the energy costs, it is necessary to predict the consequences of different pump and valve settings on the behavior of the network.

Apart from Linear and Non-linear Programming, there are other methods and approaches to design, to manage and operate a water supply network to achieve sustainability—for instance, the adoption of appropriate technology coupled with effective strategies for operation and maintenance. These strategies must include effective management models, technical support to the householders and industries, sustainable financing mechanisms, and development of reliable supply chains. All these measures must ensure the following: system working lifespan; maintenance cycle; continuity of functioning; down time for repairs; water yield and water quality.

Sustainable Development

In an unsustainable system there is insufficient maintenance of the water networks, especially in the major pipe lines in urban areas. The system deteriorates and then needs rehabilitation or renewal.

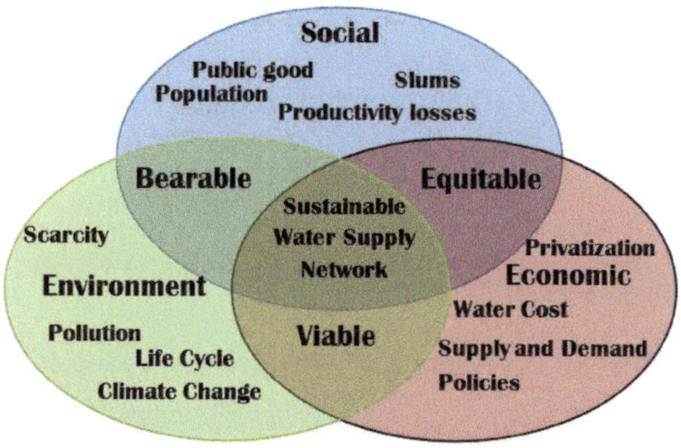

Sustainable development in an urban water network

Householders and sewage treatment plants can both make the water supply networks more efficient and sustainable. Major improvements in eco-efficiency are gained through systematic separation of rainfall and wastewater. Membrane technology can be used for recycling wastewater.

The municipal government can develop a "Municipal Water Reuse System" which is a current approach to manage the rainwater. It applies a water reuse scheme for treated wastewater, on a municipal scale, to provide non-potable water for industry, household and municipal uses. This technology consists in separating the urine fraction of sanitary wastewater, and collecting it for recycling its nutrients. The feces and graywater fraction is collected, together with organic wastes from the households, using a gravity sewer system, continuously flushed with non-potable water. The water is treated anaerobically and the biogas is used for energy production.

The sustainable water supply system is an integrated system including water intake, water utilization, wastewater discharge and treatment and water environmental protection. It requires reducing freshwater and groundwater usage in all sectors of consumption. Developing sustainable water supply systems is a growing trend, because it serves people's long-term interests. . There are several ways to re-use and recycle the water, in order to achieve long-term sustainability, such us:

- Gray water re-use and treatment: gray water is wastewater coming from baths, showers, sinks and washbasins. If this water is treated it can be used as a source of water for uses other than drinking. Depending on the type of gray water and its level of treatment, it can be re-used for irrigation and toilet flushing. According to an investigation about the impacts of domestic grey water reuse on public health, carried out by the New South Wales Health Centre in Australia in the year 2000, grey water contains less nitrogen and fecal pathogenic organisms than sewage, and the organic content of grey water decomposes more rapidly.
- Ecological treatment systems use little energy: there are many applications in gray water re-use, such us reed beds, soil treatment systems and plant filters. This process is ideal for gray water re-use, because of easier maintenance and higher removal rates of organic matter, ammonia, nitrogen and phosphorus.

Other possible approaches to scoping model's for water supply, applicable to any urban area, include the following:

- Sustainable Urban Drainage System.
- Borehole extraction.
- Intercluster groundwater flow.
- Canal and river extraction.
- Aquifer storage
- A more user-friendly indoor water use.

The "Dublin Statement on Water and Sustainable Development", mentioned above, is a good example of the new trend to overcome water supply problems. This statement, suggested by advanced economies, has come up with some principles that are of great significance to urban water supply. These are:

1. Fresh water is a finite and vulnerable resource, essential to sustain life, development and the environment.
2. Water development and management should be based on a participatory approach, involving users, planners and policy-makers at all levels.
3. Women play a central part in the provision, management and safeguarding of water. Institutional arrangements should reflect the role of women in water provision and protection

4. Water has an economic value in all its competing uses and should be recognized as an economic good. .

From these statements, developed in 1992, several policies have been created to give importance to water and to move urban water system management towards sustainable development. The Water Framework Directive by the European Commission is a good example of what has been created there out of former policies.

Future Approaches

There is great need for a more sustainable water supply systems. To achieve sustainability several factors must be tackled at the same time: climate change, rising energy cost, and rising populations. All of these factors provoke change and put pressure on management of available water resources.

An obstacle to transforming conventional water supply systems, is the amount of time needed to achieve the transformation. More specifically, transformation must be implemented by municipal legislation bodies, which always need short-term solutions too. Another obstacle to achieving sustainability in water supply systems is the insufficient practical experience with the technologies required, and the missing know-how about the organization and the transition process.

Possible ways to improve this situation is simulating of the network, implementing pilot projects, learning from the costs involved and the benefits achieved.

Water Point Mapping

Water Point Mapping (WPM) is a tool for monitoring the distribution and status of water supplies. It collects data about different aspects related to the water facility and overlays this point data with information about population and administrative boundaries. WPM helps to visualize the spatial distribution of water supply coverage and can thereby be used to highlight equity issues. The information collected provides insights into schemes' sustainability levels and management-related aspects of water points.

WPM can be used to (i) to inform the planning of investments to improve water supply coverage; (ii) to allocate resources to deliver basic services where they are most needed; (iii) to promote increased investments in the sector; and (iv) to measure progress and performance.

Relevance of Mapping

The Millennium Development Goals (MDGs) include a specific target (number 10 of Goal 7) to deal with people who do not access safe drinking water and basic sanitation.

To adequately assess peoples' access to these basic services it is vital that there is accessible, accurate and reliable data that is routinely collected and updated.

A variety of tools and techniques have been developed in recent years to collect such information. However, unless data is easily accessible and is presented in a user-friendly format, decision makers commonly do without the information. One alternative that has been designed to manage large volumes of data and to enable a user-friendly presentation is to use geo-referenced datasets, which provide a means of integration of data from different sources at any point on the globe. Within such a framework, for any specific point on the map (identified by its grid reference) detailed and accurate data of different nature can be linked in an integrated way. Mapping therefore involves the presentation of certain information in a spatial context, and this enables policy planners to identify the geographic areas and communities in which to focus their efforts for maximum impact. In all, mapping presents many benefits, such as:

- It makes easier to integrate data from different sources (surveys, censuses, satellites, etc.) and from different disciplines (social, economic, and environmental data). It also allows the switch to new units of analysis from, for example, administrative boundaries (e.g. state) to ecological boundaries (e.g. basin).
- Maps are a powerful visual tool and are more easily understood by stakeholders, particularly in developing countries.
- The spatial nature of water poverty, such as the distance to the nearest water source or the water supply infrastructure, can also be incorporated easily in a GIS database.
- The allocation of resources can be improved, since geographic targeting is more efficient and cost-effective than to launch an equally expensive universal distribution programme.
- Geo-referenced databases can be enriched by additional data as they become available; and new attributes, such as better details on water quality, can be incorporated into the data structure, ensuring that the relevance of the data is sustained over time.
- Maps can be produced at a number of different resolutions depending on their purpose and the cost of data collection. A coarse resolution or a scale too small neglects the heterogeneity within each unit and provides insufficient detail for decision making, while a fine resolution or a scale too large increases the cost of compiling, managing, and analyzing the data.

Methodology

Water Point Mapping (WPM) can be defined as an 'exercise whereby the geographical positions of all improved water points in an area are gathered in addition to management, technical and demographical information'. A handheld Global Positioning System (GPS) unit is used to record the precise location and approximate altitude of all water points audited; a digital camera is used to capture photographs of each water point in order to show the status and physical conditions; and a structured questionnaire is completed to document main characteristics, such as: location, functionality, category of water supply, water quality perception, management issues, ownership and water tariff payment. All collected data is entered into a geographical information system and correlated with available demographic, administrative, and physical data. The information is finally displayed using digital maps.

WPM seeks to serve two different purposes:

- To easily show the distribution of water points in a territory. This could enhance efficiency and transparency in local government's planning while enabling a higher degree of accountability towards population.
- To allow the definition of more reliable access indicators, which are to be constructed from the lowest geographical level with collected data.

The strength of water point maps is that they provide a clear message on who is and is not served; and particularly in rural areas, WPM can be used to highlight equity issues and schemes' functionality levels below the district level. In many developing countries, water sector ministries are undertaking reforms towards decentralization, transferring the responsibility of services management and resource allocation to local planners. This requires adequate performance monitoring frameworks backed up by accurate and reliable data at local level. WPM is increasingly thought of as a way to collect this information and display it through easy-to-use maps.

As a survey tool, WPM was originally designed and promoted by WaterAid in Malawi, although in recent years it has been carried out in different countries (Ethiopia, Ghana, Kenya, Malawi, Tanzania, Uganda, etc.) by a number of stakeholders (WaterAid, SNV, Ingeniería Sin Fronteras - ApD, Concern; etc.).

Improved and Unimproved Water Points

Main goal of WPM is to develop a comprehensive record of improved water points at a particular geographic area. The types of water points considered as improved are consistent with those accepted internationally by the Joint Monitoring Programme for Water Supply and Sanitation of WHO/UNICEF, where definition of improved is technology-based. More specifically, an improved water point is a place with some improved facilities where water is drawn for various uses such as drinking, washing and cooking.

Definitions of "improved" and "unimproved" water points	
"Improved"	**"Unimproved"**
• Piped water into dwelling, plot or yard	• Unprotected dug well
• Piped water into neighbour's plot	• Unprotected spring
• Public tap/standpipe	• Small cart with tank/drum
• Tubewell / borehole	• Tanker truck
• Protected dug well	• Surface water (river, dam, lake, etc.)
• Protected spring	• Bottled water **
• Rainwater	

**Considered as "not improved" because of concerns about the quantity of water supplied, not because of concerns over the water quality.

Water Quality and Seasonality Issues

Despite its growing importance, the issue of water quality has long been nearly absent from water coverage assessments, since few countries could afford regular water quality surveillance. Instead, it is assumed that certain types of water sources categorized as 'improved' are likely to deliver drinking water of adequate quality for basic health needs. This assumption appears over-optimistic, and improved technologies are shown not to be providing in all circumstances safe water. Similarly, health-related gains in water interventions are also dependent on service reliability; and it has been seen for example that annual health benefits attributed to the consumption of safe water are almost entirely lost if raw water is consumed over the course of few days. This draws attention to the issue of reliability and seasonality of drinking water sources.

Against this background, the Spanish NGO Ingeniería Sin Fronteras - Asociación para el Dessarrollo (ISF-ApD) complements the standard WPM campaign with two additional actions: (i) basic water quality tests are carried out using portable water testing kits (measuring pH, turbidity, chlorine, electrical conductivity and concentration of faecal coliforms); and (ii) the year-round functionality of water points is assessed by means of direct questions to users. The methodology used is ultimately designed to estimate the coverage of year-round safe improved water points of a certain territory.

Indicators for Sector Performance Monitoring

One drawback of water point maps is that they are only as accurate as the underlying data collected, so that there is a need to define reliable but simple indicators as the basis of the monitoring framework. Clearly, the core sector indicator is water supply coverage, and in rural areas, access to water is normally defined by establishing a ratio of the maximum distance and number of people served by each water point. Three different possibilities exist when defining access:

1. Calculate coverage as the standard number of people served per water point

(normally defined in the water policy) multiplied by the number of water points, regardless of distance from dwellings to the water point.

2. Calculate coverage by considering the people served by water point as the number of people living at a certain distance from it, regardless the maximum number of people to be served by a water point as defined in the corresponding water policy.

3. A case-specific approach combining both of the above conditions and applying the most restrictive one in each scenario.

In order to accurately assess the number of people served using distance as a criterion (no. 2) the spatial distribution of population is required, and this is problematic in the majority of cases. Despite the progress made with satellite free access images, many rural areas are still not covered with sufficient precision. On the other hand, due to periodic sociological census undertaken in nearly all countries, population distribution in administrative structures is usually quite well documented. Thus, first abovementioned criterion is the most commonly used.

With this information, the percentage of access to water in an area is estimated, and first indicator of access can be defined as Improved Community Water Point Density (ICWPD), which is equal to the number of ICWP per 1,000 inhabitants. A certain area would have access if its density of water points per 1,000 inhabitants is above the national threshold, and percentage of people not served in an area would be proportional to the lack of water points available compared to that threshold. However, to further improve ICWP indicator is simple since information on functionality for each water point is also collected during the survey. Non-functionality is an important obstacle to access to rural water supply, and in sub-Saharan Africa averages around 35%. Consequently, Functional Community Water Point Density (FCWPD) can be used as more accurate access indicator. If water quality and seasonality issues are taken into consideration, other 'access' indicators can be defined, as shown in following Table.

List of indicators	
Indicator	**Calculation**
• Improved community WP density (ICWPD)	• No. of improved community WPs (ICWPs), per 1,000 inhabitants
• Functional community WP density (FCWPD)	• No. of functional ICWPs (FICWPs), per 1,000 inhabitants
• Year-round functional community WP density (YRFWD)	• No. of ICWPs working at least 11 months per year, per 1,000 inhabitants
• Bacteriological acceptable WP density (BAWD)	• No. of FICWPs with an acceptable number of coliforms, per 1,000 inhabitants
• Bacteriological acceptable and year-round functional WP density (BA&YR-WD)	• No. of ICWPs working at least 11 months per year and with an acceptable number of coliforms, per 1,000 inhabitants

Combining WPM Data with Household-based Information

In addition to safe water supplies, basic sanitation and hygiene also have significant impact on household health.

Provision and use of sanitation isolates contaminated faeces from the environment, and therefore the consistent use of the facility, not its mere existence, is likely to lead to health and environmental improvements. However and in much the same way as with water supply, care is needed to define an acceptable form of excreta disposal. A wide range of technologies is used, particularly for settings where low-cost solutions are required, and relevant research conducted elsewhere concluded that all types of facility can be operated hygienically. As a result, sanitation technologies are considered as providing adequate access to sanitation as long as they are private (but not shared / public) and hygienically separate human faeces from human contact. Likewise, the addition of hygiene education is required to see health impacts materialize, in particular the basic issues of hand washing, proper disposal of faeces, and protection of drinking water. An integrated water, sanitation and hygiene (WASH) approach for data collection would consequently provide decision-makers with greater diagnostic power than focussing separately on each of these three components.

Definitions of "improved sanitation" and "unimproved" sanitation	
"Improved"	**"Unimproved"**
• Flush or pour-flush to piped sewer system	• Flush or pour-flush to elsewhere **
• Flush or pour-flush to septic tank	• Pit latrine without slab or open pit
• Flush or pour-flush to pit latrine	• Bucket
• Ventilated improved pit latrine (VIP)	• Hanging toilet or hanging latrine
• Pit latrine with slab	• No facilities or bush or field (open defecation)
• Composting toilet	• Public or shared sanitation facilities
** Excreta are flushed to the street, yard or plot, open sewer, a ditch, a drainage way or other location	

The strength of WPM is comprehensiveness with respect to the sample of water points audited, which results in the need of roughly covering the overall area of intervention. The drawback of the WPM approach, however, is that the focus is on water points, while sanitation and hygiene data need to be collected at the household. An approach for WASH data collection that combines a water points mapping with a household-based survey might overcome previous constraint. The new methodology takes the WPM as a starting point to comprehensively record all improved water points, being the water point basic sampling unit. This information can be then combined with data provided from a household survey to assess sanitation and hygienic habits. In terms of method, the approach adopted includes: (i) identification and audit of all improved water points; (ii) calculation of a sample size of households that is representative at and below the district level; and (iii) random household selection at the water point.

Applications

WPM can be used to support different types of analysis, and specific applications include among others broad strategic planning, priority setting exercises and performance monitoring. Potential users of WPM outcomes include organizations that have water poverty alleviation as their mandate, for example, at the international level, UN agencies, the World Bank, bilateral development agencies and international NGOs.

Planning at Local Level

Maps have come out a powerful tool for identifying and targeting the most deprived population, and therefore support poverty reduction initiatives. However, its potential remains underexploited. A recent study developed by WaterAid concluded that the use of this tool for better planning at the district level was still low despite the acknowledgment of its potential usefulness. Two major challenges in this regard were related to the regular updating of information and to the mechanisms to include WPM evidence in the planning process. This created a vicious circle, since the more outdated WPM data became, the less useful it was for planning purposes.

The INGO ISF-ApD and Same District Council (Tanzania) developed a methodology to update WPM data and use them to help inform planning decisions. First activity was to prepare a District Water and Sanitation priority document. A set of objective criteria were defined (using WPM information) in order to rank communities based on service level and infrastructure status. Criteria used included among others coverage, need for rehabilitation, seasonality of services, quality of water delivered, and poor management. Each criterion produced one different ranking, and this enabled an accurate prioritization of future interventions and rehabilitations in the district, a part from defining a wide range of related supporting activities. The priority document was approved and issued in September 2009 by the District Authority, and it was also used to set the priorities for all sector-related stakeholders operating within the district.

Analysis of Functionality of Water Points

WPM data can be used to analyze the durability of water points depending on different factors such as time after construction, technology, management framework, or the approach adopted in the implementation. Lessons that have been learned from this analysis might help local governments and other agencies to address the sustainability challenge.The Water Point Data Exchange (WPDx), launched in 2015, is a global platform for sharing water point data collected by governments, non-profit organizations, researchers, and others.

Advocacy

Maps are also a powerful tool for advocacy, since they provide clear messages and communicate complex information quickly and accurately. In addition, local governments have used mapping evidences to lobby for additional funding, and citizens have used them to pressure their representatives for better services.

At the same time, diverse initiatives are underway to display maps via web, as an attempt to foster accountability and disseminate WPM outputs to a broader public. For example, the h2.0 initiative of UN Habitat and Google, the Water Point Mapper developed by WaterAid, or the support made by the Water and Sanitation program to the establishment of WatSan portals in many African countries.

Researchers at ODI found that Water Mapping could improve accountability in service delivery. Accountability could be achieved through the creation of Community Maps and the public availability of mapping knowledge which empower the community.

Monitoring the Sector

WPM is an adequate tool to support monitoring and evaluation of the sector. Water poverty follows a highly heterogeneous pattern, widely varying between and within different geographic and administrative units; and mapping permits a feasible visualization of such heterogeneity. In addition, maps are well suited to illustrate spatial change over time.

Non-revenue Water

Non revenue water (NRW) is water that has been produced and is "lost" before it reaches the customer. Losses can be real losses (through leaks, sometimes also referred to as physical losses) or apparent losses (for example through theft or metering inaccuracies). High levels of NRW are detrimental to the financial viability of water utilities, as well to the quality of water itself. NRW is typically measured as the volume of water "lost" as a share of net water produced. However, it is sometimes also expressed as the volume of water "lost" per km of water distribution network per day.

NRW Components and Audits

The International Water Association (IWA) has developed a detailed methodology to assess the various components of NRW. Accordingly, NRW has the following components:

- Unbilled authorized consumption
- Apparent losses (water theft and metering inaccuracies)
- Real losses (from transmission mains, storage facilities, distribution mains or service connections)

In many utilities the exact breakdown of NRW components and sub-components is simply not known, making it difficult to decide about the best course of action to reduce NRW. Metering of water use at the level of production (wells, bulk water supply), at key points in the distribution network and for consumers is essential to estimate levels of NRW.

In most developed countries, there are no or very limited apparent losses. For developing countries the World Bank has estimated that, on average, apparent losses - in particular theft through illegal connections - account for about 40% of NRW. In some cities, apparent losses can be higher than real losses. Reducing apparent losses from illegal connections is often beyond what a utility can achieve by itself, because it requires a high level of political support. Illegal connections are often in slums, which means that their regularization in some cases particularly affects the poor. A water audit is a key tool to assess the breakdown of NRW and to develop a program for NRW reduction. Often a distinction is made between unvalidated and validated water audits. Unvalidated water audits are desktop studies that include many estimates and their results can have an error range for real losses of +/- 50% or more. Its main value is to identify where it is necessary to reduce the uncertainty of the water audit through validation. Validating water audits is a complex process that involves testing of production water meters, testing of a representative random sample of customer meters, eliminating systematic errors created through the billing process and validating the number of illegal connections through aerial mapping, field surveys or cross-references between various existing databases. In developing countries it is rare to find utilities that have undertaken validated water audits, and even in developed countries they are not systematically used. The American Water Works Association (AWWA) has developed Water Audit Software which allows utilities to rate the overall degree of validity of their water audit data. Guidance on loss control planning is given based upon the credibility of the data and the measure of losses displayed by the water audit.

NRW is sometimes also referred to as unaccounted-for water (UFW). While the two terms are similar, they are not identical, since non-revenue water includes authorized unbilled consumption (e.g. for firefighting or, in some countries, for use by religious institutions) while unaccounted-for water excludes it.

NRW Indicators

The most commonly used indicator to measure NRW is the percentage of NRW as a share of water produced. While this indicator is easy to understand and indeed has been widely used, it has increasingly been recognized that it is not an appropriate indicator to benchmark NRW levels between utilities or even to monitor changes over time. When losses in terms of absolute volume are constant the percentage of NRW varies greatly with total water use, i.e. if water use increases and the volume of losses remains constant the percentage of NRW declines. This problem can be eliminated by measuring NRW not as a share, but in terms of absolute losses per connection per day, as recommended by the International Water Association (IWA). Nevertheless, the use of percentage figures to compare levels of NRW remains common despite its shortcomings. The International Benchmarking Network for Water and Sanitation recommends to use different indicators (percentage, losses per connection or losses per km of network) together. Losses per kilometer of network are more appropriate to benchmark real losses, while losses per connection are more appropriate to benchmark apparent losses.

The concept of NRW as an indicator to compare *real* losses of water utilities has been criticized as flawed, particularly because real losses depend to some extent on factors largely outside the control of the utility, such as topography, age of network, length of network per connection and water use per capita. As an alternative indicator for the measurement of real losses an Infrastructure Leakage Index (ILI) has been developed. The ILI is defined as the ratio of Current Annual Real Losses (CARL) to Unavoidable Annual Real Losses (UARL).

Overview of NRW levels

Expressed as a Share of Produced Water

The following percentages indicate the share of NRW in total water produced:

- Singapore 5% (UFW)
- Denmark 6%
- Netherlands 6%
- Germany 7% (2005)
- Japan 7% (2007)
- Eastern Manila, Philippines 11% (2011), down from 63% in 1997
- Tunisia 18% (2004)
- England and Wales 19% (2005)
- MWA, Bangkok 25% (2012)
- France 26% (2005)
- Dhaka, Bangladesh 29% (2010)
- Italy 29% (2005)
- Chile 34% (2006)
- Eastern Jakarta, Indonesia 39% (2011)
- Amman, Jordan 34% (2010)

- Mexico 51% (2003)
- Western Jakarta, Indonesia 39% (2011), down from 57% in 1998
- Bauchi state, Nigeria 70%
- Yerevan, Armenia 72% (1999)
- Lagos, Nigeria 96% (pre-2003)

Expressed in Cubic Meters Per Network Length

The following figures are expressed in cubic meters per kilometer of distribution network per day:

- Netherlands 1.5
- Denmark 1.6
- Germany (towns) 0.7 - 2.4
- Germany (large cities) 2.4 - 5
- Australia 4.4
- Malmö, Sweden 5
- California Water Service Company 6
- Portugal 7
- England and Wales 10
- Helsinki 18
- Penn American Water 19
- Russia 20 (2006)
- Stockholm 21
- Scotland 21.3
- Illinois American Water 26
- Ireland 29
- Brazil 42 (2006)
- China 52 (2006)
- Bucharest 350 in 2000 and 176 in 2007

These levels are given per km of network, not per connection.

Benefits of NRW Reduction

The World Bank has estimated the total cost of NRW to utilities worldwide at US$14 billion per year. Reducing by half the current levels of losses in developing countries, where relative losses are highest, could generate an estimated US$2.9 billion in cash and serve an additional 90 million people.

Benefits of NRW reduction, in particular of leakage reduction, include:

- financial gains from increased water sales or reduced water production, including possibly the delay of costly capacity expansion;
- increased knowledge about the distribution system;
- increased firefighting capability due to increased pressure;
- reduced property damage;
- reduced risk of contamination;
- More stabilized water pressure throughout the system.

Leakage reduction may also be an opportunity to improve relations with the public and employees. A leak detection program may be highly visible, encouraging people to think about water conservation. The reduction of commercial losses, while politically and socially challenging, can also improve relations with the public, since some consumers may be reluctant to pay their water bills knowing that many others use services without being billed or being underbilled.

In the specific context of the United States NRW reduction can also mean reduced legal liability and reduced insurance payments.

Programs to Reduce NRW

Reducing NRW is a complex process. While some programs have been successful, there are many pitfalls.

Successful Programs

In the following cities high levels of non-revenue water have been susbtantially reduced:

- Dolphin Coast (iLembe), South Africa, 30% in 1999 to 16% in 2003 by the private utility Siza Water Company;
- Istanbul, Turkey, from more than 50% prior to 1994 to 34% in 2000 by the public utility ISKI;
- Jamshedpur, India, from an estimated 36% in 2005 to 10% in 2009 by the private utility Jamshedpur Utilities and Services Company;

- East Manila, Philippines, from 63% in 1997 to 16% in 2009 by the private utility Manila Water;
- Ouagadougou and other cities in Burkina Faso, by the public utility Office National de l'Eau et de l'Assainissement (ONEA) which achieved a level of 16% in 2008;
- Paranaguá, Brazil, from 58% in 2000 to 38% in 2006 by a private utility;
- Phnom Penh, Cambodia, from 72% in 1993 to 6% in 2008 by the public utility Phnom Penh Water Supply Authority (PPWSA);
- Five municipalities in Rio de Janeiro State (Prolagos), Brazil, from 60% in 2000 to 36% in 2006 by a private utility;
- Rabat, Morocco, from 32% in 2002 to 19% in 2008 by the private utility REDAL;
- Cities in Senegal, from 32% in 1996 to 20% in 2006 by the private utility Senegalaise des Eaux;
- Tangiers, Morocco from 41% in 2002 to 21% in 2008 by the private utility Amendis.
- 8 districts in Johor State, Malaysia, from 38% in 2004 to 29% in 2011 by the private utility Ranhill Utilities

These successes were achieved by both public and private utilities, in every continent, in emerging countries as well as very poor countries, in large cities and smaller towns. All required a long-term commitment by utility management and the government - local or national - over a period of at least four years.

Pitfalls of Programs

Many programs to reduce NRW have failed to achieve their objectives, sometimes from the onset and sometimes only in the long run. Often they focus on real losses without sufficient attention being paid to apparent losses. If programs achieve an initial reduction in NRW levels, they often increase again over the years to the same or even higher levels than before the program. Both apparent and real losses have a natural tendency to increase if nothing is done: more leakage will occur, there will be more defective meters, and information on customers and networks will become more outdated. In order to sustain NRW at low levels, investments in fixing leaks and replacing meters are insufficient in the best case and ineffective in the worst case. To achieve permanent results, management procedures related to a utility's organization, precedures and human resources have to be changed. Additionally the implementation of an Intelligent Pressure management system is an efficient approach to reduce the total real losses in the long term. It is one of the most basic and lucrative forms of optimizing a system and generally provides fast investment paybacks.

According to a study by the World Bank some of the reasons why NRW levels in developing countries have not been reduced significantly are the following.

Physical loss reduction is an ongoing, meticulous activity with few supporters among the following:

- Politicians: there is no "ribbon cutting" involved.
- Engineers: it is more "fun" to design treatment plants than to fix pipes buried under the road.
- Technicians and field staff: detection is done primarily at night, and pipe repairs often require working in hazardous traffic conditions.
- Managers: it needs time, constant dedication, staff, and up-front funding.

Nor is the reduction of commercial losses very popular among the following:

- Politicians: unpopular decisions might have to be made (disconnection of illegal consumers or customers who don't pay).
- Meter readers: fraudulent practices might generate a substantial additional income.
- Field staff: working on detecting illegal connections or on suspending service for those who don't pay their bills is unpopular and can even be dangerous.
- Managers: it is easier to close any revenue gap by just spending less on asset rehabilitation (letting the system slowly deteriorate) or asking the government for more money.

  —World Bank, The Challenge of Reducing Non-Revenue Water in Developing Countries

Another source quotes the seven most frequent reasons for failure of NRW reduction programs as follows:

- Poor design
- Diagnoses based on preconceptions rather than experimentation
- Partial implementation
- Failure to mobilize the necessary human and financial resources
- Lack of coordination between the components of the program
- Underestimation of the difficulties
- Underestimation of the time factor

Optimal Level of NRW

There is some debate as to what is an economically optimal level of leakage or, speaking more broadly, of NRW. From a financial or economic point of view it is not appropriate to try to reduce NRW to the lowest possible level, because the marginal cost of reducing NRW increases once the cheaper options have been exploited. Once the marginal cost of reducing NRW exceeds the marginal benefits or water savings, an economic optimum has been achieved. Benefits should be measured through reduced production costs if reduction of NRW results in lower water production, through the avoided costs of additional supply capacity if the system is close to the limit of its capacity and demand is growing, or through the value of water sold if reduction of NRW results in additional water sales. The latter can be done by valuing water through water tariffs (financial value) or through the willingness to pay by customers (economic value). There are fewer financial incentives for a utility to reduce NRW if water production is cheap, if there is no or little metering (so that revenues thus are independent of actual consumption), or if volumetric tariffs are low.

In the United Kingdom the assessment of economic levels of leakage has a long history. The first national study on the topic was published in 1980 setting down a methodology for the assessment of economic leakage levels. This led to the implementation of sectors (District Metered Areas) in most water companies in the UK. The findings were reported in a major national research program in 1994. As a result of a drought in 1995/96 a number of companies initiated major leakage management programmes based on economic assessments. The situation in other parts of the world is quite different from the UK. Particularly in developing countries sectorisation is very rare and proactive leakage control limited. The benefits of pressure management are not widely appreciated and there is generally no assessment of the economic level of leakage.

From a public health and drinking water quality point of view it is being argued that the level of real water losses should be as low as possible, independently of economic or financial considerations, in order to minimize the risk of drinking water contamination in the distribution network.

The World Bank recommends that NRW should be “less than 25%”, while the Chilean water regulator SISS has determined a NRW level of 15% as optimal in its model of an efficient water company that it uses to benchmark service providers. In England and Wales NRW stands at 19% or 149 liter/property/day.

In the United States the American Water Works Association’s (AWWA) Water Loss Control Committee recommended in 2009 that water utilities conduct annual water audits as a standard business practice. AWWA recommends that water utilities should track volumes of apparent and real losses and the annual cost impacts of these losses. Utilities should then seek to control excessive losses to levels that are economic for the water utility. In 1999 the California Urban Water Conservation Council identified a 10 percent benchmark for non-revenue water.

WASH

Celebrating the opening of a water supply project in Isabel Province, Solomon Islands

WASH (also called Watsan, WaSH or WSH) stands for "Water, Sanitation and Hygiene" - several interrelated public health issues that are of particular interest to international development programs. Affordable access to WASH is a key public health issue, especially in developing countries.

Several international development agencies have identified WASH as an area with significant potential to improve health, life expectancy, student learning, gender equality, and other important issues of international development.

Access to WASH, in particular safe water, supply adequate sanitation, and proper hygiene education, can reduce illness and death, and also impact poverty reduction and socio-economic development. Lack of sanitation contributes to approximately 700,000 child deaths every year due to diarrhea. Chronic diarrhea can have a negative effect on child development (both physical and cognitive). In addition, lack of WASH facilities can prevent students from attending school, impose a burden on women and diminish productivity.

In developping countries, access to WASH is challenging and in 2015 the World Health Organization (WHO) estimates that still "2.5 billion people – more than one third of the global population – live without basic sanitation facilities" while 750 million people still lack access to safe and clean drinking water.

Background

Distributing jerrycans to help people store clean drinking water in the Philippines

The concept of WASH, groups together water, sanitation, and hygiene because the impact of deficiencies in each area overlap strongly. Addressing these deficiencies together, can achieve a strong positive impact on public health.

Women line up at a bore hole to fill their containers with water (Labuje IDP camp, Kitgum, Kitgum District, Northern Region of Uganda)

The UN's Millennium Development Goals included improvement of WASH services in Target 7.C: "Halve, by 2015, the proportion of the population without sustainable access to safe drinking water and basic sanitation." This has been replaced by the Sustainable Development Goals where Target 6 includes: "Ensure availability and sustainable management of water and sanitation for all".

Access to WASH, in particular safe water, adequate sanitation, and proper hygiene education, can reduce illness and death, and also impact poverty reduction and socio-economic development. Lack of sanitation contributes to approximately 700,000 child deaths every year due to diarrhea. Chronic diarrhea can have a negative effect on child development (both physical and cognitive). In addition, lack of WASH facilities can prevent students from attending school, impose a burden on women and diminish productivity.

Although access to sanitation has been improving over the past decades, the World Health Organization (WHO) estimates that still "2.5 billion people – more than one third of the global population – live without basic sanitation facilities". In 2015, 750 million people lacked access to safe, clean drinking water and approximately 2,300 people die every day from diarrhea.

The United Nation's International Year of Sanitation in 2008 has helped to increase attention for funding of sanitation in WASH programs of many donors. For example, the Bill and Melinda Gates Foundation, has increased their funding levels of sanitation projects since 2009, with a strong focus on reuse of excreta.

Activities

Awareness Raising

Awareness raising for the importance of WASH is regularly carried out by various organizations through their publications and activities on certain special days of the year (United Nations international observance days), namely: World Water Day for water (22 March), World Toilet Day for sanitation (26 November) and Global Handwashing Day for hygiene (15 October).

Neglected Tropical Diseases

Water, sanitation and hygiene interventions help to prevent many neglected tropical diseases (NTDs), for example soil-transmitted helminthiasis. An integrated approach to NTDs and WASH benefits both sectors and the communities they are aiming to serve. This is especially true in areas that are endemic with more than one NTD.

A map has been created to help identify areas with high levels of infection with the WASH-impacted NTDs and low levels of rural improved water and sanitation coverage. In addition, WASH practitioners can use the manual "WASH and the Neglected Tropical Diseases: A Manual for WASH Implementers" to target, implement, and monitor WASH program impact on the NTDs.

In August 2015, the World Health Organization (WHO) unveiled a global strategy and action plan to integrate WASH with other public health interventions in order to accelerate elimination of NTDs. The plan aims to intensify control or eliminate certain NTDs in specific regions by 2020. It refers to the NTD roadmap milestones that included for example eradication of dracunculiasis by 2015 and of yaws by 2020, elimination of trachoma and lymphatic filariasis as public health problems by 2020, intensified control of dengue, schistosomiasis and soil-transmitted helminthiases. The plan consists of four strategic objectives: Improving awareness of benefits of joint WASH and NTD actions; monitoring WASH and NTD actions to track progress; strengthening evidence of how to deliver effective WASH interventions; and planning, delivering and evaluating WASH and NTD programmes with involvement of all stakeholders. The aim is to use synergies between WASH and NTD programmes.

WASH in Schools

More than half of all primary schools in the developing countries with available data do not have adequate water facilities and nearly two thirds lack adequate sanitation. Even where facilities exist, they are often in poor condition. WASH in schools, sometimes called SWASH or WinS, significantly reduces hygiene related disease, increases student attendance and contributes to dignity and gender equality. WASH in schools contributes to healthy, safe and secure school environments that can protect children from health hazards, abuse and exclusion. It also enables children to become agents of change for improving water, sanitation and hygiene practices in their families and communities.

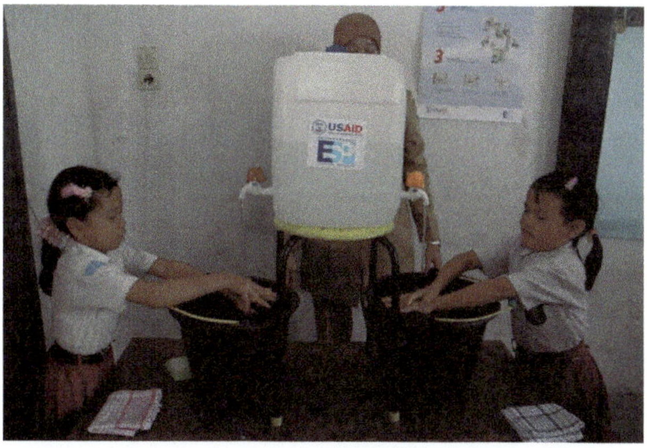

Pupils in Medan, Indonesia, practice handwash in class

Reasons for missing or poorly maintained water and sanitation facilities at schools in developing countries include lacking inter-sectoral collaboration; lacking cooperation between schools, communities and different levels of government; as well as a lack in leadership and accountability. Methods to overcome these problems can include on a policy level: broadening the focus of the education sector, establishing a systematic quality assurance system, distributing and using funds wisely. And on a practical level it is recommended to: Have a clear and systematic mobilization strategy, support the education sector to strengthen intersectoral partnerships, establish a constant monitoring system which is located within the education sector, educate the educators and partner with the school management.

Supervised daily group handwashing in schools can be an effective strategy for building hygiene habits, with the potential to lead to positive health and education outcomes for children.

Strong cultural taboos around menstruation, which are present in many societies, coupled with a lack of Menstrual Hygiene Management services in schools, results in girls staying away from school during menstruation.

Challenges

Urban Slums

Part of the reason for slow progress in sanitation may be due to the "urbanization of poverty," as poverty is increasingly concentrated in urban areas. Migration to urban areas, resulting in denser clusters of poverty, poses a challenge for sanitation infrastructures that were not originally designed to serve so many households, if they existed at all.

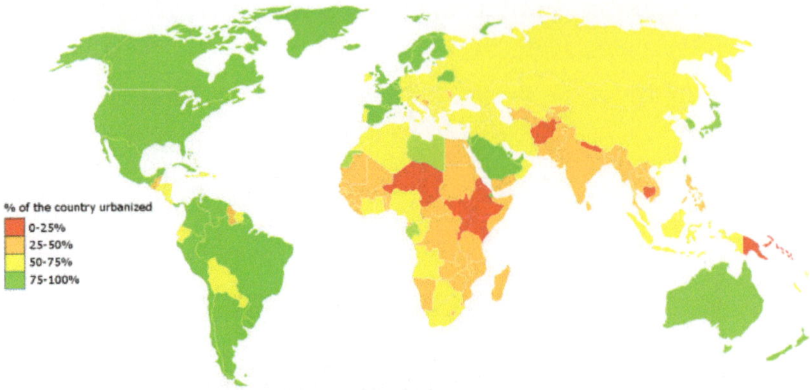

Rapid growth of urbanisation

Waiting in line for two hours to collect water from a standpipe.

As poverty becomes more concentrated in urban areas, one increasingly common phenomenon is the expansion of urban slums. Often built illegally in response to a lack of more permanent housing, slums have a specific set of problems associated with them. For instance, the lack of property rights and instability associated with a slum dwelling may mean that the resident would not be willing to invest in WASH services for a building that may not survive a storm, or from which she may be evicted. In addition, "New urban areas may be very heterogeneous—both ethnically and in terms of wealth distribution. They may face a constant influx of new migrants." Such heterogeneity may make it difficult to coordinate efforts to build and maintain a shared sanitation system for slum neighborhoods.

For example, Oxfam is helping to provide 1 million litres of water each day in Mingkamen, but the demand is such that people must wait up to two hours in line to fill their buckets. Sometimes fights break out at the water point because everyone is waiting so long.

National WASH Plans and Monitoring

A WHO report found that only one-third of the countries surveyed have national WASH plans that are being properly implemented, funded and regularly reviewed. In most countries monitoring was inconsistent and there were critical gaps. Reliable data is essential to inform policy decision, to monitoring and evaluate outcomes, and to identify those who do not have access to WASH. Many countries have WASH monitoring frameworks in place, but most of the data reported was inconsistent, weakening evaluation and outcome data analysis.

Failures of WASH Systems

National government mapping and monitoring efforts as well as post-project monitoring by NGOs or researchers, have identified the failure of water supply systems (also known as water points, wells, or boreholes) and sanitation systems (one part of sanitation systems are the toilets).

History

The abbreviation WASH was used from the year 1988 onwards as an acronym for the "Water and Sanitation for *Health*" Project of the United States Agency for International Development. At that time, the letter "H" stood for "health", not "hygiene". Similarly, in Zambia the term WASHE was used in a report in 1987 and stood for "Water Sanitation Health Education". An even older USAID "WASH project report" dates back to as early as 1981.

From about 2001 onwards, international organizations active in the area of water supply and sanitation advocacy, such as the Water Supply and Sanitation Collaborative Council and the International Water and Sanitation Centre (IRC) in the Netherlands began to use "WASH" as an umbrella term for water, sanitation and hygiene. "WASH" has since then been broadly adopted as a handy acronym for water, sanitation and hygiene in the international development context. The term "WatSan" was also used for a while, especially in the emergency response sector such as with IFRC and UNHCR, but has not proven as popular as WASH.

The term "water" in the acronym WASH is generally understood to refer to water supply only, not e.g. to integrated water resources management (IWRM) or water resource management in agriculture.

References

- Zhang, Stephen X, Babovic, Vladan (2012). "A real options approach to the design and architecture of water supply systems using innovative water technologies under uncertainty". Journal of Hydroinformatics. 14 (1): 13–29. doi:10.2166/hydro.2011.078
- Programme, United Nations Human Settlements (2003). Facing the slum challenge : global report on human settlements, 2003 (PDF) (Repr. ed.). London: Earthscan Publications. p. xxvi. ISBN 1-84407-037-9. Retrieved 30 April 2015

- "Neglected Tropical Diseases and WASH Index Map". Water, Sanitation & Hygiene for Neglected Tropical Disease Online Manual Resource. Retrieved 9 July 2015
- Harvey, P. A., and Reed, R. A. (2007). "Community-managed water supplies in Africa: sustainable or dispensable?" Community Development Journal, 42(3), 365-378
- UN-Water Global Analysis and Assessment of Sanitation and Drinking Water (2014). Investing in Water and Sanitation: Increasing Access, Reducing Inequalities (GLAAS 2014 Report). World Health Organization. p. iv. ISBN 978 92 4 150808 7. Retrieved 30 April 2015
- UNHCR Division of Operational Services (2008). A Guidance for UNHCR Field Operations on Water and Sanitation Services. Geneva, Switzerland. Retrieved 11 March 2016
- Dogan Altinbilek:Water Management in Istanbul, International Journal of Water Resources Development, Vol. 22, No. 2, 241-253, June 2006
- Superintendencia de Servicios Sanitarios (SISS) (2007). "Informe de gestión del sector sanitario 2006" (PDF) (in Spanish). Retrieved 2008-02-13. , p. 63
- Johnston, E. A.; Teague, Jordan; Graham, Jay P. (2015-06-11). "Challenges and opportunities associated with neglected tropical disease and water, sanitation and hygiene intersectoral integration programs". BMC Public Health. 15 (1): 547. ISSN 1471-2458. PMC 4464235. PMID 26062691. doi:10.1186/s12889-015-1838-7

2

Water Supply: Infrastructures and Transportation

Aqueducts are canals, tunnels and other structures that are used to transport water. Water kiosk, spring supply, water tower, water tank, rainwater harvesting and water well are some of the water supply infrastructures. This chapter is an overview of the subject matter incorporating all the major aspects of water supply infrastructures.

Aqueduct (Water Supply)

An aqueduct is a watercourse constructed to convey water. In modern engineering, the term *aqueduct* is used for any system of pipes, ditches, canals, tunnels, and other structures used for this purpose. The term *aqueduct* also often refers specifically to a bridge on an artificial watercourse. The word is derived from the Latin *aqua* ("water") and *ducere* ("to lead"). Aqueducts were used in ancient Greece, ancient Egypt, and ancient Rome. In modern times, the largest aqueducts of all have been built in the United States to supply the country's biggest cities. The simplest aqueducts are small ditches cut into the earth. Much larger channels may be used in modern aqueducts. Aqueducts sometimes run for some or all of their path through tunnels constructed underground. Modern aqueducts may also use pipelines. Historically, agricultural societies have constructed aqueducts to irrigate crops and supply large cities with drinking water.

Ancient Aqueducts

Although particularly associated with the Romans, aqueducts were devised much earlier in Greece and the Near East and Indian subcontinent, where peoples such as the Egyptians and Harappans built sophisticated irrigation systems. Roman-style aqueducts were used as early as the 7th century BC, when the Assyrians built an 80 km long limestone aqueduct, which included a 10 m high section to cross a 300 m wide valley, to carry water to their capital city, Nineveh.

India

The Indian subcontinent is believed to have some of the earliest aqueducts. Evidence can be found at the sites of present-day Hampi, Karnataka. The massive aqueducts near river Tungabhadra supplying irrigation water were once 15 miles (24 km) long. The waterways supplied water to royal bath tubs.

Ancient Indian aqueduct in Hampi

Oman

In Oman from the Iron Age, in Salut, Bat, and other sites, a system of underground aqueducts called falaj or qanāts were constructed, a series of well-like vertical shafts, connected by gently sloping horizontal tunnels.

There are three types of falaj:

- Daudi with underground aqueducts
- Ghaili requiring a dam to collect the water
- Aini whose source is a water spring

These enabled large scale agriculture to flourish in a dry land environment.

Persia

In Persia from early times a system of underground aqueducts called qanāts were constructed, a series of well-like vertical shafts, connected by gently sloping tunnels. This technique:

- taps into subterranean water in a manner that delivers water to the surface without need for pumping. The water drains relying on gravity, with the destination lower than the source, which is typically an upland aquifer.
- allows water to be transported long distances in hot dry climates without losing a large proportion of the source water to seepage and evaporation.

Petra, Jordan

Throughout Petra, Jordan, the Nabataean engineers took advantage of every natural spring and every winter downpour to channel water where it was needed. They constructed aqueducts and piping systems that allowed water to flow across mountains, through gorges and into the temples, homes, and gardens of Petra's citizens. Walking through the Siq, one can easily spot the remains of channels that directed water to the city center, as well as durable retention dams that kept powerful flood waters at bay.

Greece

On the island of Samos, the Tunnel of Eupalinos was built during the reign of Polycrates (538-522 BC). It is considered an underground aqueduct and brought fresh water to Pythagoreion for roughly a thousand years.

Roman

The multiple arches of the Pont du Gard, in Roman Gaul. Its lower tiers carry a road across the river, and the upper tiers support an aqueduct conduit that carried water to Nimes

Roman aqueducts were built in all parts of the Roman Empire, from Germany to Africa, and especially in the city of Rome, where they totaled over 415 kilometres (258 mi). The aqueducts supplied fresh water to public baths and for drinking water, in large cities across the empire, and set a standard of engineering that was not surpassed for more than a thousand years. Bridges, built in stone with multiple arches, were a distinctive feature of Roman aqueducts and hence the term *aqueduct* is often applied specifically to a bridge for carrying water.

South America

Near the Peruvian town of Nazca, an ancient pre-Columbian system of aqueducts called Puquios were built and are still in use today. They were made of intricately placed stones, a construction material widely used by the Nazca culture. The time period in which they were constructed is still debated, but some evidence supports circa A.D. 540–552, in response to drought periods in the region.

Underground aqueducts of Cantalloc, Nazca, Peru

The Guayabo National Monument of Costa Rica, a park covering the largest archaeological site in the country, contains a system of aqueducts. The complex network of uncovered and covered aqueducts still functions well. The aqueducts are constructed from rounded river stones, which are mostly made of volcanic rock. The civilization that constructed the aqueduct system remains a mystery to archaeologists; it is suspected that Guayabo's aqueducts sat at a point of ancient cultural confluence between Aztecs, Mayans, and Incas.

North America

When Europeans saw the Aztec capital Tenochtitlán, early in the 16th century, the city was watered by two aqueducts. One of these, Chapultepec Aqueduct, built circa 1420, was rebuilt by the Spanish almost three hundred years later. Originally tracing part of its path over now-gone Lake Texcoco, only a fragment remains in Mexico City today.

Sri Lanka

Extensive usage of elaborate aqueducts have been found to have been used in ancient Sri Lanka.The best example is the Yoda Ela or Jaya Ganga, an 87 kilometres (54 mi) long water canal carrying excess water between two artificial reservoirs with a gradient of 10 to 20 cm per kilometer during the fifth century AD. However, the ancient engineering methods in calculating the exact elevation between the two reservoirs and the exact gradient of the canal to such fine precision had been lost with the fall of the civilization in 13th Century.

Modern Aqueducts

In modern times, the largest aqueducts of all have been built in the United States to supply the country's biggest cities. The Catskill Aqueduct carries water to New York City over a distance of 120 miles (190 km), but is dwarfed by aqueducts in the far west of the country, most notably the Colorado River Aqueduct, which supplies the Los Angeles area with water from the Colorado River nearly 400 km to the east and the 701.5 mi (1,129 km) California Aqueduct, which runs from the Sacramento-San Joaquin River Delta to Lake Perris. The Central Arizona Project is the largest and most expensive aqueduct constructed in the United States. It stretches 336 miles from its source near Parker, Arizona to the metropolitan areas of Phoenix and Tucson. An aqueduct in New Zealand, "the Oamaru Borough Race" was constructed in the late 19th century to deliver water (and water-power) about 50 km from the Waitaki River at Kurow to the coastal town of Oamaru.

In Spain, the Tagus-Segura Water Transfer system of aqueducts opened in 1979 and transports water 286 kilometres (178 mi) from north to south.

In China, the South–North Water Transfer Project aims to connect the Yangtze River basin to Beijing through three separate systems. The project will reuse part of the Grand Canal of China.

Design

Open Channels

The simplest aqueducts are small ditches cut into the earth. Much larger channels may be used in modern aqueducts, for instance the Central Arizona Project uses 7.3 m (24 ft) wide channels. A major factor in the design of all open channels is its gradient. A higher gradient allows a smaller channel to carry the same amount of water as a larger channel with a lower gradient, but increases the potential of the water to damage the aqueduct's structure. A typical Roman aqueduct had a gradient of about 1:4800.

Underground Tunnels

Aqueducts sometimes run for some or all of their path through tunnels constructed underground. A version of this common in North Africa and Central Asia that has vertical wells at regular intervals is called a qanat. One historic example found in Syria, the Qanat Firaun, extends over 100 kilometers.

Pipes

The Los Angeles Aqueduct

Modern aqueducts may also make extensive use of pipelines. Pipelines are useful for transporting water over long distances when it needs to move over hills, or where open channels are poor choices due to considerations of evaporation, freezing, pollution, or environmental impact. They can also be used to carry treated water.

Uses

Historically, agricultural societies have constructed aqueducts to irrigate crops. Archimedes invented the water screw to raise water for use in irrigation of croplands.

Another use for aqueducts is to supply large cities with drinking water. It also help drought prone areas with water supplies. Some of the Roman aqueducts still supply water to Rome today. In California, United States, three large aqueducts supply water over hundreds of miles to the Los Angeles area. Two are from the Owens River area, and a third is from the Colorado River.

In modern civil engineering projects, detailed study and analysis of open-channel flow is commonly required to support flood control, irrigation systems, and large water supply systems when an aqueduct rather than a pipeline is the preferred solution.

In the past, aqueducts often had channels made of earth or other porous materials but significant amounts of water are lost through such unlined aqueducts. As water gets increasingly scarce, these canals are being lined with concrete, polymers, or impermeable soil. In some cases, a new aqueduct is built alongside the old one because it cannot be shut down during construction.

Water Kiosk

Water kiosks are booths for the sale of tap water. They are common in many countries of Sub-Saharan Africa. Water kiosks exist, among other countries, in Cameroon, Haiti, Kenya, Malawi, Tanzania and Zambia.

Description

Water kiosks typically have four taps on the outside and faucets inside, operated by a kiosk attendant. They typically receive treated water from utilities through a piped distribution network. Where water supply in the network is intermittent, kiosks sometimes also have a water storage tank. In rural areas, water for kiosks can sometimes come directly from a well, spring, stream or lake after treatment. An example for a kiosk supplied from a lake is a village near Gona dam in Kenya.

Management and Sustainability

Kiosks can be operated by employees of utilities, by self-employed operators under contract with utilities or water committees consisting of volunteers. Kiosk operators also sell other goods at the kiosk to increase their meager income. A water kiosk can serve between 500 and 3,000 people. Water is typically carried home from the kiosk in buckets of 20 liters. The sale price can be a flat rate per household or, more typically, a price per bucket which is advertised at the kiosk.

Making water kiosks commercially viable is more difficult where population density is low and where there are alternative often low-quality, free water sources such as shallow wells, ponds or streams. Low awareness of the health benefits of clean water can exacerbate these problems. In those conditions, kiosks are at greater risk of failing. Involving communities in deciding about the location of kiosks, their opening hours and the choice of the kiosk operators increases the likelihood of the kiosks being accepted and functioning well. If operators have a contract with the utility, regular supervision is important to ensure that the contract stipulations concerning cleanliness of the premises, prices charged and opening hours are respected.

Experiences

Kenya:Regularization of Informal Kiosks in Kibera

In Kibera, Nairobi's largest slum, water kiosks exist since the 1970s. Kiosks are privately owned and the owners financed the construction of the kiosks and the pipes to the water mains. Water is supplied by the Nairobi utility, but is often not paid for by the kiosk owners. In 1998 there were about 650 kiosks in Kibera. Although two thirds of the kiosks have water reservoirs, often water is not available due to supply interruptions. In 2003, when a new water law was passed, the government threatened to shut down kiosks that were not properly registered. As a result, kiosk owners formed an association and engaged in a dialogue with the government to defuse the situation by paying arrears and being officially recognized.

Zambia:Successful Country-wide Program

In Zambia, where kiosks were introduced in 2006, there were about 300 kiosks in 2008 serving 500,000 people. According to the World Bank and the German development agency GIZ they are a success. More than half the kiosk operators in Zambia are women. The utilities have set up so-called "peri-urban units" that control the kiosk operators every two weeks and replace operators that are in breach of their contract. The tariff charged is 1 Euro Cent per 20 liter.

Haiti:Success in the Capital, Mixed Results in Towns

In Port-au-Prince, Haiti, water kiosks have been operated successfully in low-income neighborhoods through water committees since the mid-1990s. The water committees use about a third of their revenues to pay for water provided by the utility and one third for other operating costs. The remaining third is used for community activities. However, water kiosks in other towns have been abandoned. Some have struggled, because nearby households receive water through unmetered house connections for a flat rate and resell water to neighbors, thus depriving the kiosks of clients.

Malawi:Failure and a New Attempt

In Lilongwe, the capital of Malawi, water kiosks have failed at one moment. 49 kiosks were set up in four neighborhoods and each kiosk was managed by a committee of three people. However, some committees misappropriated funds, while some community members got free water through illegal piped connections directly to their homes. Some kiosks charged flat rates per month independently of consumption, but users often failed to pay. When some committees did not pay the utility, the latter threatened to shut down 27 water kiosks. The NGO WaterAid conducted a survey to assess the situation and recommended that the Malawi Water Authority establishes a designated kiosk unit to liaise with the committees running the kiosks.

Other Examples

In Douala, Cameroon, and Dar es-Salaam, Tanzania, the tariff for water sold at kiosks is about 3 Euro Cent per 20 liter. This is still much lower than the prices charged by water vendors which the poor had to pay prior to the construction of the kiosks.

Spring Supply

A spring supply is a provision of piped mains water to a number of consumers direct from a natural spring. Spring supplies are therefore a source of groundwater, which in most instances has fewer micro-organisms (e.g. coliform bacteria and protozoa such as *Giardia* and *Cryptosporidium*) and chemical contaminants than a supply from surface water. The point at which the groundwater reaches the surface is prone to contamination, so must be protected using a structure called a spring box. This is often surrounded by a fence to keep animals out, with other common features being a ditch on the uphill side, an overflow pipe and a well fitting lid. Spring supplies can range from single property supplies that are privately owned, to large supplies that are managed by water companies and serve entire communities. As with any water supply, a spring supply may need to be treated in order to bring it up to drinking water standards. The method for doing this will vary according to the contaminant, but can include sand filters, pH balancing units and ultraviolet light.

Countries

In the United Kingdom, over half a million people live or work in a premises that relies on a private water supply such as a spring. The Drinking Water Inspectorate

(DWI) of England and Wales produces an annual report on the quality of private water supplies.

Examples

The hamlet of Pinchinthorpe in Teesside, England, for example, has a spring supply.

Water Tower

A water tower is an elevated structure supporting a water tank constructed at a height sufficient to pressurize a water supply system for the distribution of potable water, and to provide emergency storage for fire protection. In some places, the term standpipe is used interchangeably to refer to a water tower, especially one with tall and narrow proportions. Water towers often operate in conjunction with underground or surface service reservoirs, which store treated water close to where it will be used. Other types of water towers may only store raw (non-potable) water for fire protection or industrial purposes, and may not necessarily be connected to a public water supply.

Water towers are able to supply water even during power outages, because they rely on hydrostatic pressure produced by elevation of water (due to gravity) to push the water into domestic and industrial water distribution systems; however, they cannot supply the water for a long time without power, because a pump is typically required to refill the tower. A water tower also serves as a reservoir to help with water needs during peak usage times. The water level in the tower typically falls during the peak usage hours of the day, and then a pump fills it back up during the night. This process also keeps the water from freezing in cold weather, since the tower is constantly being drained and refilled.

History

Although the use of elevated water storage tanks has existed since ancient times in various forms, the modern use of water towers for pressurized public water systems developed during the mid-19th century, as steam-pumping became more common, and better pipes that could handle higher pressures were developed. In the United Kingdom, standpipes consisted of tall, exposed, n-shaped pipes, used for pressure relief and to provide a fixed elevation for steam-driven pumping engines which tended to produce a pulsing flow, while the pressurized water distribution system required constant pressure. Standpipes also provided a convenient fixed location to measure flow rates. Designers typically enclosed the riser pipes in decorative masonry or wooden structures. By the late 19th-Century, standpipes grew to include storage tanks to meet the ever-increasing demands of growing cities.

Many early water towers are now considered historically significant and have been included in various heritage listings around the world. Some are converted to apartments or exclusive penthouses. In certain areas, such as New York City in the United States, smaller water towers are constructed for individual buildings. In California and some other states, domestic water towers enclosed by siding (tankhouses) were once built (1850s–1930s) to supply individual homes; windmills pumped water from hand-dug wells up into the tank in New York.

Design and Construction

A variety of materials can be used to construct a typical water tower; steel and reinforced or prestressed concrete are most often used (with wood, fiberglass, or brick also in use), incorporating an interior coating to protect the water from any effects from the lining material. The reservoir in the tower may be spherical, cylindrical, or an ellipsoid, with a minimum height of approximately 6 metres (20 ft) and a minimum of 4 m (13 ft) in diameter. A standard water tower typically has a height of approximately 40 m (130 ft).

Pressurization occurs through the hydrostatic pressure of the elevation of water; for every 10.20 centimetres (4.016 in) of elevation, it produces 1 kilopascal (0.145 psi) of pressure. 30 m (98.43 ft) of elevation produces roughly 300 kPa (43.511 psi), which is enough pressure to operate and provide for most domestic water pressure and distribution system requirements.

Shooter's Hill water tower is a local landmark in London, United Kingdom.
Water towers are common around London suburbs.

The height of the tower provides the pressure for the water supply system, and it may be supplemented with a pump. The volume of the reservoir and diameter of the piping provide and sustain flow rate. However, relying on a pump to provide pressure is expensive; to keep up with varying demand, the pump would have to be sized to meet peak demands. During periods of low demand, jockey pumps are used to meet these lower water flow requirements. The water tower reduces the need for electrical consumption of cycling pumps and thus the need for an expensive pump control system, as this system would have to be sized sufficiently to give the same pressure at high flow rates.

Very high volumes and flow rates are needed when fighting fires. With a water tower present, pumps can be sized for average demand, not peak demand; the water tower can provide water pressure during the day and pumps will refill the water tower when demands are lower.

Using wireless sensor networks to monitor water levels inside the tower allows municipalities to automatically monitor and control pumps without installing and maintaining expensive data cables.

Decoration

Louisville Water Tower, one of the few remaining standpipe water towers in the United States. It was completed in 1860.

Water towers can be surrounded by ornate coverings including fancy brickwork, a large ivy-covered trellis or they can be simply painted. Some city water towers have the name of the city painted in large letters on the roof, as a navigational aid to aviators and motorists. Sometimes the decoration can be humorous. An example of this are water towers built side by side, labeled *HOT* and *COLD*. Cities in the United States possessing side-by-side water towers labeled HOT and COLD include Granger, Iowa; Canton, Kansas;Pratt, Kansas and St. Clair, Missouri (Eveleth, Minnesota at one time had two such towers, but no longer does). When a third water tower was built next to the Okemah, Oklahoma set of Hot and Cold towers, the town briefly considered naming it "Running", but eventually decided to use "Home of Woody Guthrie". *The House in the Clouds* in Thorpeness, located in the English county of Suffolk, was built to resemble a house in order to disguise the eyesore, whilst the lower floors were used for accommodation. When the town was connected to the mains water supply, the water tower was dismantled and converted to additional living space.

The *House in the Clouds* in Thorpeness functioned as the town's water tower from 1923 until 1977.

Sapp Bros. truck stops use a water tower with a handle and spout – looking like a coffee pot – as the company logo. Many of their facilities have decorated actual water towers (presumably non-functional) on-site.

Tower with local high school mascot, a Tiger (Centerville, Texas).

The first and original "Mushroom" – *Svampen* in Swedish – was built in Örebro in Sweden in the early 1950s and later copies were built around the world including Saudi-Arabia and Kuwait.

Many small towns in the United States use their water towers to advertise local tourism, their local high school sports teams, or other locally notable facts. Since the water tower is sometimes the highest point in the town, antennas, public address systems, cameras and tornado warning sirens are sometimes placed on them as well.

Many water towers serve manufacturing and other commercial facilities. These are often decorated with the name of the company that the water tower serves.

Worldwide

The technology dates to at least the 19th century, and for a long time New York City required that all buildings higher than six stories be equipped with a rooftop water tower. Two companies in New York build water towers, both of which are family businesses in operation since the 19th century.

The original water tower builders were barrel makers who expanded their craft to meet a modern need as buildings in the city grew taller in height. Even today, no sealant is used to hold the water in. The wooden walls of the water tower are held together with steel cables or straps, but water leaks through the gaps when first filled. As the water saturates the wood, it swells, the gaps close and become impermeable.

Rooftop water towers on apartment buildings on East 57th Street in New York City. The structures seen here illustrate three architectural approaches to incorporating these tanks in the design of a building. From left to right, a fully enclosed and ornately decorated brick structure, a simple unadorned roofless brick structure hiding most of the tank but revealing the top of the tank, and a simple utilitarian structure that makes no effort to hide the tanks or otherwise incorporate them into the design of the building.

The rooftop water towers store 25,000 to 50,000 litres (5,500 to 11,000 imp gal) of water until it is needed in the building below. The upper portion of water is skimmed off the top for everyday use while the water in the bottom of the tower is held in reserve to fight fire. When the water drops below a certain level, a pressure switch, level switch or float valve will activate a pump or open a public water line to refill the water tower.

Architects and builders have taken varied approaches to incorporating water towers into the design of their buildings. On many large commercial buildings, water towers are completely hidden behind an extension of the facade of the building. For cosmetic reasons, apartment buildings often enclose their tanks in rooftop structures, either simple unadorned rooftop boxes, or ornately decorated structures intended to enhance the visual appeal of the building. Many buildings, however, leave their water towers in plain view atop utilitarian framework structures.

Water towers are common in India, where the electricity supply is erratic in most places.

The mushroom-shaped concrete water tower of Roihuvuori in Helsinki, Finland was built in the 1970s. It is 52 metres (171 ft) high and can hold around 12,000 cubic metres (420,000 cu ft) of water.

If the pumps fail (such as during a power outage), then water pressure will be lost, causing potential public health concerns. Many U.S. states require a "boil-water advisory" to be issued if water pressure drops below 20 pounds per square inch (140 kPa). This advisory presumes that the lower pressure might allow pathogens to enter the system.

Water towers are often regarded as monuments of civil engineering. Some are converted to serve modern purposes, as for example, the Wieża Ciśnień (Wrocław water tower) in Wrocław, Poland which is today a restaurant complex. Others have been converted to residential use.

Historically, railroads that used steam locomotives required a means of replenishing the locomotive's tenders. Water towers were common along the railroad. The tenders were usually replenished by water cranes, which were fed by a water tower.

Some water towers are also used as observation towers, and some restaurants, such as the Goldbergturm in Sindelfingen, Germany, or the second of the three Kuwait Towers, in the City-State of Kuwait. It is also common to use water towers as the location of transmission mechanisms in the UHF range with small power, for instance for closed rural broadcasting service, amateur radio, or cellular telephone service.

In hilly regions, local topography can be substituted for structures to elevate the tanks. These tanks are often nothing more than concrete cisterns terraced into the sides of local hills or mountains, but function identically to the traditional water tower. The tops of these tanks can be landscaped or used as park space, if desired.

Spheres and Spheroids

The Chicago Bridge and Iron Company have built many of the water spheres and spheroids found in the United States. The website *World's Tallest Water Sphere* describes the distinction between a water sphere and water spheroid thus:

Eindhoven Water Towers

A water sphere is a type of water tower that has a large sphere at the top of its post. The sphere looks like a golf ball sitting on a tee or a round lollipop. A cross section of a sphere in any direction (east-west, north-south, or top-bottom) is a perfect circle. A water spheroid looks like a water sphere, but the top is wider than it is tall. A spheroid looks like a round pillow that is somewhat flattened. A cross section of a spheroid in two directions (east-west or north-south) is an ellipse, but in only one direction (top-bottom) is it a perfect circle. Both spheres and spheroids are special-case ellipsoids: spheres have symmetry in 3 directions, spheroids have symmetry in 2 directions. Scalene ellipsoids have 3 unequal length axes and three unequal cross sections.

The Union Watersphere is a water tower topped with a sphere-shaped water tank in Union, New Jersey and characterized as the *World's Tallest Water Sphere*. A Star Ledger article suggested a water tower in Erwin, North Carolina completed in early 2012, 219.75 ft (66.98 m) tall and holding 500,000 US gallons (1,900 m^3), had become the World's Tallest Water Sphere. However photographs of the Erwin water tower revealed the new tower to be a water spheroid. The water tower in Braman, Oklahoma, built by the Kaw Nation and completed in 2010, is 220.6 ft (67.2 m) tall and can hold 350,000 US gallons (1,300 m^3). Slightly taller than the Union Watersphere, it is also a spheroid. Another tower in Oklahoma, built in 1986 and billed as the *"largest water tower in the country"*, is 218 ft (66 m) tall, can hold 500,000 US gallons (1,900 m³), and is located in Edmond.

The *Earthoid*, a perfectly spherical tank located in Germantown, Maryland is 100 ft (30 m) tall and holds 2,000,000 US gallons (7,600 m³) of water. The name is taken from it being painted to resemble a globe of the world. The golf ball-shaped tank of the water tower at Gonzales, California is supported by three tubular legs and reaches about 125 ft (38 m) high. The Watertoren (or Water Towers) in Eindhoven, Netherlands contain three spherical tanks, each 10 m (33 ft) in diameter and capable of holding 500 cubic metres (130,000 US gal) of water, on three 43.45 m (142.6 ft) spires were completed in 1970.

Alternatives

Mechelen-Zuid water tower, one of the tallest in the world.

Alternatives to water towers are simple pumps mounted on top of the water pipes to increase the water pressure. This new approach is more straightforward, but also more subject to potential public health risks; if the pumps fail, then loss of water pressure may result in entry of contaminants into the water system. Most large water utilities do not use this approach, given the potential risks.

Tallest

Tower	Year	Country	Town	Pinnacle height	Remarks
Swisscom-Sendeturm St. Chrischona	1984	Switzerland	St. Chrischona	250 m	

Tower	Year	Country	Town	Pinnacle height	Remarks
Kuwait Towers, Tower A	1979	Kuwait	Kuwait City	187 m	
Kuwait Towers, Tower B	1979	Kuwait	Kuwait City	146 m	
Waldenburg TV Tower	1959	Germany	Waldenburg	145 m	partially guyed tower consisting of water tower and antenna mast guyed to the ground as pinnacle, antenna mast was dismantled in 2008
Mechelen-Zuid water tower	1978	Belgium	Mechelen	143 m	combined water and telecommunications tower
Ginosa Water Tower					

Water Tank

Water tank used in Municipal Water Treatment Plant Houston Texas.

Water Tank Towers, featured in Kuwait City.

A water tank is a container for storing water. Water tanks are used to provide storage of water for use in many applications, drinking water, irrigation agriculture, fire suppression, agricultural farming, both for plants and livestock, chemical manufacturing, food preparation as well as many other uses. Water tank parameters include the general design of the tank, and choice of construction materials, linings. Various materials are used for making a water tank: plastics (polyethylene, polypropylene), fiberglass, concrete, stone, steel (welded or bolted, carbon, or stainless). Earthen pots also function as water storages. Water tanks are an efficient way to help developing countries to store clean water.

Water tank from 1876 in Hanover, Germany

History

Stone water tank in a castle courtyard, Saxony, Germany.

An old-fashioned water tank near Twentynine Palms, California.

Throughout history, wood, ceramic and stone tanks have been used as water tanks. These containers were all naturally occurring and some man made and a few of these tanks are still in service. The Indus Valley Civilization (3000–1500 BC) made use of granaries and water tanks. Medieval castles needed water tanks for the defenders to withstand a siege. A wooden water tank found at the Año Nuevo State Reserve (California) was restored to functionality after being found completely overgrown with ivy. It had been built in 1884.

Types

Chemical contact tank of FDA and NSF polyethylene construction, allows for retention time for chemical treatment chemicals to "contact" (chemically treat) with product water.

The taanka is used in Rajasthan as a traditional form of rainwater harvesting.

Ground water tank, made of lined carbon steel, may receive water from a water well or from surface water, allowing a large volume of water to be placed in inventory and used during peak demand cycles.

Elevated water tank, also known as a water tower, will create a pressure at the ground-level outlet of 1 kPa per 10.2 cm or 1 psi per 2.31 feet of elevation. Thus a tank elevated to 20 metres creates about 200 kPa and a tank elevated to 70 feet creates about 30 psi of discharge pressure, sufficient for most domestic and industrial requirements.

Vertical cylindrical dome top tanks may hold from 200 litres or fifty gallons to several million gallons. Horizontal cylindrical tanks are typically used for transport because their low-profile creates a low center of gravity helping to maintain equilibrium for the transport vehicle, trailer or truck.

A Hydro-pneumatic tank is typically a horizontal pressurized storage tank. Pressurizing this reservoir of water creates a surge free delivery of stored water into the distribution system.

Design

By design a water tank or container should do no harm to the water. Water is susceptible to a number of ambient negative influences, including bacteria, viruses, algae, changes in pH, and accumulation of minerals, accumulated gas. The contamination can come from a variety of origins including piping, tank construction materials, animal and bird feces, mineral and gas intrusion. A correctly designed water tank works to address and mitigate these negative effects.

A safety based news article linked copper poisoning as originating from a plastic tank. The article indicated that rainwater was collected and stored in a plastic tank and that the tank did nothing to mitigate the low pH. The water was then brought into homes with copper piping, the copper was released by the high acid rainwater and caused poisoning in humans. It is important to note that since the plastic tank is an inert container, it has no effect on the incoming water. Good practice would be to analyze any water source periodically and treat accordingly, in this case the collected acid rain should be analyzed, and pH adjusted before being brought into a domestic water supply system.

The release of copper due to acidic water is monitored may be accomplished with a variety of technology, beginning with pH strips and going to more sophisticated pH monitors, indicate pH which when acidic or caustic, some with output communication capabilities. There is no "linkage" between the plastic tank and copper poisoning, a solution to the problem is easy, monitor 'stored rainwater' with 'swimming pool strips' cheap and available at, swimming pool supply outlets. If the water is too acidic, contact state/county/local health officials to obtain advice and precise solutions and pH limits and guidelines as to what should be used to treat rainwater to be used as domestic drinking water.

Articles and specifications for water tank applications and design considerations, the AWWA (American Water Works Association) provides details as required by many states to complete a certification process to insure the quality of water being consumed.

The American Water Works Association is a reservoir of water tank knowledge; the association provides specifications for a variety of water storage tank applications as well as design. The AWWA's site provides scientific resources with which the reader will be able to develop an informed perspective on which to make decisions regarding their water tank requirements.

Water Well

A water well is an excavation or structure created in the ground by digging, driving, boring, or drilling to access groundwater in underground aquifers. The well water is drawn by a pump, or using containers, such as buckets, that are raised mechanically or by hand. Wells were first constructed at least eight thousand years ago and historically vary in construction from a simple scoop in the sediment of a dry watercourse to the stepwells of India, the qanats of Iran, and the shadoofs and sakiehs of India. Placing a lining in the well shaft helps create stability and linings of wood or wickerwork date back at least as far as the Iron Age.

Wells have been traditionally sunk by hand digging as is the case in rural developing areas. These wells are inexpensive and low-tech as they use mostly manual labour and the structure can be lined with brick or stone as the excavation proceeds. A more modern method called caissoning uses pre-cast reinforced concrete well rings that are lowered into the hole. Driven wells can be created in unconsolidated material with a well hole structure, which consists of a hardened drive point and a screen of perforated pipe, after which a pump is installed to collect the water. Deeper wells can be excavated by hand drilling methods or machine drilling, using a bit in a borehole. Drilled wells are usually cased with a factory-made pipe composed of steel or plastic. Drilled wells can access water at much greater depths than dug wells.

A dug well in a village in Faryab Province, Afghanistan.

Two broad classes of well are shallow or unconfined wells completed within the uppermost saturated aquifer at that location, and deep or confined wells, sunk through an impermeable stratum into an aquifer beneath. A collector well can be constructed adjacent to a freshwater lake or stream with water percolating through the intervening material. The site of a well can be selected by a hydrogeologist, or groundwater surveyor. Water may be pumped or hand drawn. Impurities from the surface can easily reach shallow sources and contamination of the supply by pathogens or chemical contaminants needs to be avoided. Well water typically contains more minerals in solution than surface water and may require treatment before being potable. Soil salination can occur as the water table falls and the surrounding soil begins to dry out. Another environmental problem is the potential for methane to seep into the water.

Types

Leather bucket used for the water well.

A dug well, large shadoof (well sweep), and old barn in Markowa, Poland.

Well, Historical Village, Bhaini Sahib, Ludhiana, Punjab, India.

Dug Wells

View into a hand-dug well cased with concrete rings; Ouelessebougou, Mali.

A dug well in a village in Kerala, India.

Until recent centuries, all artificial wells were pumpless hand-dug wells of varying degrees of sophistication, and they remain a very important source of potable water in some rural developing areas where they are routinely dug and used today. Their indispensability has produced a number of literary references, literal and figurative, to them, including the reference to the incident of Jesus meeting a woman at Jacob's well (John 4:6) in the bible and the "Ding Dong Bell" nursery rhyme about a cat in a well.

Hand-dug wells are excavations with diameters large enough to accommodate one or more people with shovels digging down to below the water table. The excavation is braced horizontally to avoid landslide or erosion endangering the people digging. They can be lined with laid stones or brick; extending this lining upwards above the ground surface to form a wall around the well serves to reduce both contamination and injuries by falling into the well. A more modern method called caissoning uses reinforced concrete or plain concrete pre-cast well rings that are lowered into the hole. A well-digging team digs under a cutting ring and the well column slowly sinks into the aquifer, whilst protecting the team from collapse of the well bore.

Hand-dug wells are inexpensive and low tech (compared to drilling) as they use mostly manual labour to access groundwater in rural locations in developing countries. They may be built with a high degree of community participation, or by local entrepreneurs who specialize in hand-dug wells. They have been successfully excavated to 60 metres (200 ft).

They have low operational and maintenance costs, in part because water can be extracted by hand bailing, without a pump. The water is often coming from an aquifer or groundwater, and can be easily deepened, which may be necessary if the ground water level drops, by telescoping the lining further down into the aquifer. The yield of existing hand dug wells may be improved by deepening or introducing vertical tunnels or perforated pipes.

Drawbacks to hand-dug wells are numerous. It can be impractical to hand dig wells in areas where hard rock is present, and they can be time-consuming to dig and line even in favourable areas. Because they exploit shallow aquifers, the well may be susceptible to yield fluctuations and possible contamination from surface water, including sewage. Hand dug well construction generally requires the use of a well trained construction team, and the capital investment for equipment such as concrete ring moulds, heavy lifting equipment, well shaft formwork, motorized de-watering pumps, and fuel can be large for people in developing countries. Construction of hand dug wells can be dangerous due to collapse of the well bore, falling objects and asphyxiation, including from dewatering pump exhaust fumes.

Woodingdean well, hand-dug between 1858 and 1862, is claimed to be the world's deepest hand-dug well at 392 metres (1,285 ft). The Big Well in Greensburg, Kansas is billed as the world's largest hand-dug well, at 109 feet (33 m) deep and 32 feet (9.8 m) in diameter. However, the *Well of Joseph* in the Cairo Citadel at 280 feet (85 m) deep and the Pozzo di S. Patrizio (St. Patrick's Well) built in 1527 in Orvieto, Italy, at 61 metres (200 ft) deep by 13 metres (43 ft) wide are both larger by volume.

Driven Wells

Driven wells may be very simply created in unconsolidated material with a *well hole structure*, which consists of a hardened drive point and a screen (perforated pipe). The point is simply hammered into the ground, usually with a tripod and *driver*, with pipe sections added as needed. A driver is a weighted pipe that slides over the pipe being driven and is repeatedly dropped on it. When groundwater is encountered, the well is washed of sediment and a pump installed.

Drilled Wells

Drilled wells are typically created using either top-head rotary style, table rotary, or cable tool drilling machines, all of which use drilling stems that are turned to create a cutting action in the formation, hence the term *drilling*.

Drilled wells can be excavated by simple hand drilling methods (augering, sludging, jetting, driving, hand percussion) or machine drilling (rotary, percussion, down the hole hammer). Deeprock rotary drilling method is most common. Rotary can be used in 90% of formation types.

Drilled wells can get water from a much deeper level than dug wells can—often down to several hundred metres.

Cable tool water well drilling rig in Kimball, West Virginia, USA.

Water well drilling in Ein Hemed, near Jerusalem circa 1964.

Drilled wells with electric pumps are used throughout the world, typically in rural or sparsely populated areas, though many urban areas are supplied partly by municipal wells. Most shallow well drilling machines are mounted on large trucks, trailers, or tracked vehicle carriages. Water wells typically range from 3 to 18 metres (10–60 ft) deep, but in some areas can go deeper than 900 metres (3,000 ft).

Rotary drilling machines use a segmented steel drilling string, typically made up of 6 metres (20 ft) sections of galvanized steel tubing that are threaded together, with a bit or other drilling device at the bottom end. Some rotary drilling machines are designed to install (by driving or drilling) a steel casing into the well in conjunction with the drilling of the actual bore hole. Air and/or water is used as a circulation fluid to displace cuttings and cool bits during the drilling. Another form of rotary style drilling, termed *mud rotary*, makes use of a specially made mud, or drilling fluid, which is constantly being altered during the drill so that it can consistently create enough hydraulic pressure to hold the side walls of the bore hole open, regardless of the presence of a casing in the well. Typically, boreholes drilled into solid rock are not cased until after the drilling process is completed, regardless of the machinery used.

The oldest form of drilling machinery is the cable tool, still used today. Specifically designed to raise and lower a bit into the bore hole, the *spudding* of the drill causes the bit to be raised and dropped onto the bottom of the hole, and the design of the cable causes the bit to twist at approximately $\frac{1}{4}$ revolution per drop, thereby creating a drilling action. Unlike rotary drilling, cable tool drilling requires the drilling action to be stopped so that the bore hole can be bailed or emptied of drilled cuttings.

Drilled wells are usually cased with a factory-made pipe, typically steel (in air rotary or cable tool drilling) or plastic/PVC (in mud rotary wells, also present in wells drilled into solid rock). The casing is constructed by welding, either chemically or thermally, segments of casing together. If the casing is installed during the drilling, most drills will drive the casing into the ground as the bore hole advances, while some newer machines will actually allow for the casing to be rotated and drilled into the formation in a similar manner as the bit advancing just below. PVC or plastic is typically welded and then lowered into the drilled well, vertically stacked with their ends nested and either glued or splined together. The sections of casing are usually 6 metres (20 ft) or more in length, and 6 to 12 in (15 to 30 cm) in diameter, depending on the intended use of the well and local groundwater conditions.

Surface contamination of wells in the United States is typically controlled by the use of a *surface seal*. A large hole is drilled to a predetermined depth or to a confining formation (clay or bedrock, for example), and then a smaller hole for the well is completed from that point forward. The well is typically cased from the surface down into the smaller hole with a casing that is the same diameter as that hole. The annular space between the large bore hole and the smaller casing is filled with bentonite clay, concrete, or other sealant material. This creates an impermeable seal from the surface to the next confining layer that keeps contaminants from traveling down the outer sidewalls of the casing or borehole and into the aquifer. In addition, wells are typically capped with either an engineered well cap or seal that vents air through a screen into the well, but keeps insects, small animals, and unauthorized persons from accessing the well.

At the bottom of wells, based on formation, a screening device, filter pack, slotted casing, or open bore hole is left to allow the flow of water into the well. Constructed screens are typically used in unconsolidated formations (sands, gravels, etc.), allowing water and a percentage of the formation to pass through the screen. Allowing some material to pass through creates a large area filter out of the rest of the formation, as the amount of material present to pass into the well slowly decreases and is removed from the well. Rock wells are typically cased with a PVC liner/casing and screen or slotted casing at the bottom, this is mostly present just to keep rocks from entering the pump assembly. Some wells utilize a *filter pack* method, where an undersized screen or slotted casing is placed inside the well and a filter medium is packed around the screen, between the screen and the borehole or casing. This allows the water to be filtered of unwanted materials before entering the well and pumping zone.

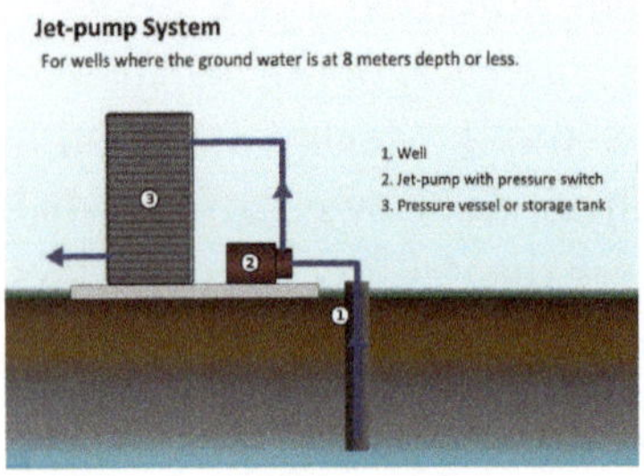

An automated water well system powered by a jet-pump.

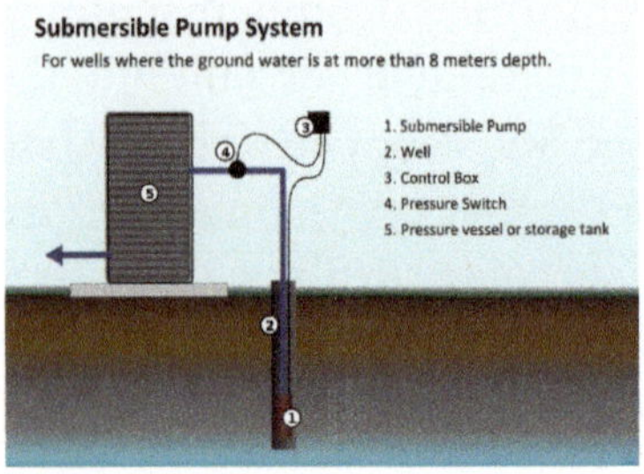

An automated water well system powered by a submersible pump.

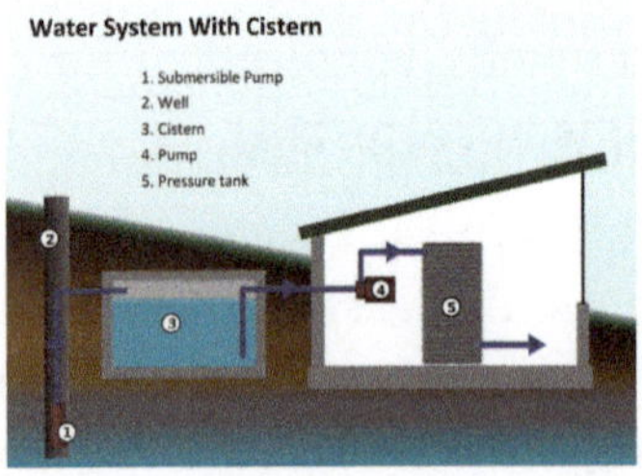

A water well system with a cistern.

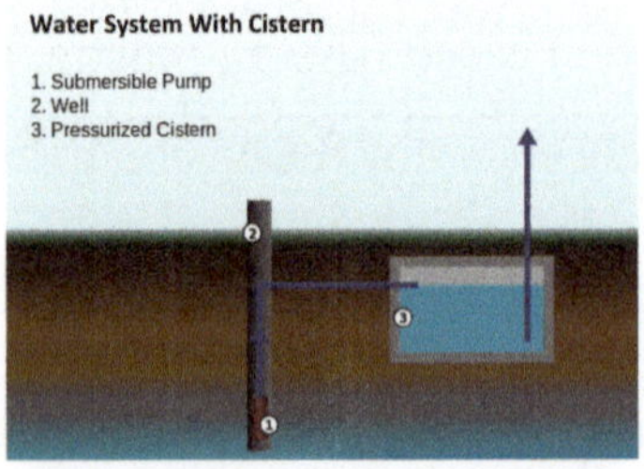

A water well system with a pressurized cistern.

Classification

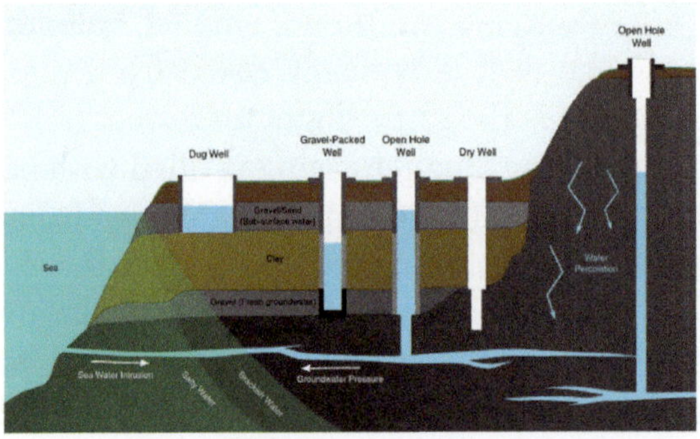

Water well types

There are two broad classes of drilled-well types, based on the type of aquifer the well is in:

- *Shallow* or *unconfined wells* are completed in the uppermost saturated aquifer at that location (the upper unconfined aquifer).
- *Deep* or *confined wells* are sunk through an impermeable stratum into an aquifer that is sandwiched between two impermeable strata (aquitards or aquicludes). The majority of deep aquifers are classified as artesian because the hydraulic head in a confined well is higher than the level of the top of the aquifer. If the hydraulic head in a confined well is higher than the land surface it is a "flowing" artesian well (named after Artois in France).

A special type of water well may be constructed adjacent to freshwater lakes or streams.

Ctommonly called a collector well but sometimes referred to by the trade name Ranney well or Ranney collector, this type of well involves sinking a caisson vertically below the top of the aquifer and then advancing lateral collectors out of the caisson and beneath the surface water body. Pumping from within the caisson induces infiltration of water from the surface water body into the aquifer, where it is collected by the collector well laterals and conveyed into the caisson where it can be pumped to the ground surface.

Two additional broad classes of well types may be distinguished, based on the use of the well:

- *production* or *pumping wells*, are large diameter (greater than 15 cm in diameter) cased (metal, plastic, or concrete) water wells, constructed for extracting water from the aquifer by a pump (if the well is not artesian).
- *monitoring wells* or *piezometers*, are often smaller diameter wells used to monitor the hydraulic head or sample the groundwater for chemical constituents. Piezometers are monitoring wells completed over a very short section of aquifer. Monitoring wells can also be completed at multiple levels, allowing discrete samples or measurements to be made at different vertical elevations at the same map location.

A well constructed for pumping groundwater can be used passively as a monitoring well and a small diameter well can be pumped, but this distinction by use is common.

Siting

Before excavation, information about the geology, water table depth, seasonal fluctuations, recharge area and rate must be found. This work is typically done by a hydrogeologist, or a groundwater surveyor using a variety of tools including electro-seismic surveying, any available information from nearby wells, geologic maps, and sometimes geophysical imaging.

Contamination

Man cleaning a well in Yaoundé, Cameroon

Shallow pumping wells can often supply drinking water at a very low cost. However, impurities from the surface easily reach shallow sources, which leads to a greater risk of contamination for these wells compared to deeper wells. Contaminated wells can lead to the spread of various waterborne diseases. Dug and driven wells are relatively easy to contaminate; for instance, most dug wells are unreliable in the majority of the United States.

Hand pump to pump water from a well in a village near Chennai in India, where the well water might be polluted by nearby pit latrines.

Pathogens

Most of the bacteria, viruses, parasites, and fungi that contaminate well water comes from fecal material from humans and other animals, for example from on-site sanitation systems (such as pit latrines and septic tanks). Common bacterial contaminants include *E. coli*, *Salmonella*, *Shigella*, and *Campylobacter jejuni*. Common viral contaminants include *norovirus*, *sapovirus*, *rotavirus*, enteroviruses, and hepatitis A and E. Parasites include *Giardia lamblia*, *Cryptosporidium*, *Cyclospora cayetanensis*, and microsporidia.

Chemical Contamination

Chemical contamination is a common problem with groundwater. Nitrates from sewage, sewage sludge or fertilizer are a particular problem for babies and young children. Pollutant chemicals include pesticides and volatile organic compounds from gasoline, dry-cleaning, the fuel additive methyl tert-butyl ether (MTBE), and perchlorate from rocket fuel, airbag inflators, and other artificial and natural sources.

Several minerals are also contaminants, including lead leached from brass fittings or old lead pipes, chromium VI from electroplating and other sources, naturally occurring arsenic, radon, and uranium—all of which can cause cancer—and naturally occurring fluoride, which is desirable in low quantities to prevent tooth decay, but can cause dental fluorosis in higher concentrations.

Some chemicals are commonly present in water wells at levels that are not toxic, but can cause other problems. Calcium and magnesium cause what is known as hard water, which can precipitate and clog pipes or burn out water heaters. Iron and manganese can appear as dark flecks that stain clothing and plumbing, and can promote the growth of iron and manganese bacteria that can form slimy black colonies that clog pipes.

Prevention

The quality of the well water can be significantly increased by lining the well, sealing the well head, fitting a self-priming hand pump, constructing an apron, ensuring the area is kept clean and free from stagnant water and animals, moving sources of contamination (pit latrines, garbage pits, on-site sewer systems) and carrying out hygiene education. The well should be cleaned with 1% chlorine solution after construction and periodically every 6 months.

Well holes should be covered to prevent loose debris, animals, animal excrement, and wind-blown foreign matter from falling into the hole and decomposing. The cover should be able to be in place at all times, including when drawing water from the well. A suspended roof over an open hole helps to some degree, but ideally the cover should be tight fitting and fully enclosing, with only a screened air vent.

Minimum distances and soil percolation requirements between sewage disposal sites and water wells need to be observed. Rules regarding the design and installation of private and municipal septic systems take all these factors into account so that nearby drinking water sources are protected.

Education of the general population in society also plays an important role in protecting drinking water.

Mitigation

Cleanup of contaminated groundwater tends to be very costly. Effective remediation of groundwater is generally very difficult. Contamination of groundwater from surface and subsurface sources can usually be dramatically reduced by correctly centering the casing during construction and filling the casing annulus with an appropriate sealing material. The sealing material (grout) should be placed from immediately above the production zone back to surface, because, in the absence of a correctly constructed casing seal, contaminated fluid can travel into the well through the casing annulus. Centering devices are important (usually 1 per length of casing or at maximum intervals of 9 m) to ensure that the grouted annular space is of even thickness. Upon the construction of a new test well, it is considered best practice to invest in a complete battery of chemical and biological tests on the well water in question. Point-of-use treatment is available for individual properties and treatment plants are often constructed for municipal water supplies that suffer from contamination. Most of these treatment methods involve the filtration of the contaminants of concern, and additional protection may be garnered by installing well-casing screens only at depths where contamination is not present.

Well water for personal use is often filtered with reverse osmosis water processors; this process can remove very small particles. A simple, effective way of killing microorganisms is to bring the water to a full boil for one to three minutes, depending on location. A household well contaminated by microorganisms can initially be treated by shock chlorination using bleach, generating concentrations hundreds of times greater than found in community water systems; however, this will not fix any structural problems that led to the contamination and generally requires some expertise and testing for effective application.

After the filtration process, it is common to implement an Ultraviolet (UV) system to kill pathogens in the water. UV light effects the DNA of the pathogen by UV-C photons breaking through the cell wall. UV disinfection has been gaining popularity in the past decades as it is a chemical free method of water treatment.

Environmental Problems

A risk with the placement of water wells is soil salination which occurs when the water table of the soil begins to drop and salt begins to accumulate as the soil begins to dry out. Another environmental problem that is very prevalent in water well drilling is the potential for methane to seep through.

Soil Salination

The potential for soil salination is a large risk when choosing the placement of water wells. Soil salination is caused when the water table of the soil drops over time and salt begins to accumulate. In turn, the increased amount of salt begins to dry the soil out. This is a very detrimental problem because the increased level of salt in the soil can result in the degradation of soil and can be very harmful to vegetation.

Methane

Methane, an asphyxiant, is a chemical compound that is the main component of natural gas. When methane is introduced into a confined space, it displaces oxygen, reducing oxygen concentration to a level low enough to pose a threat to humans and other aerobic organisms but still high enough for a risk of spontaneous or externally caused explosion. This potential for explosion is what poses such a danger in regards to the drilling and placement of water wells.

Low levels of methane in drinking water are not considered toxic. When methane seeps into a water supply, it is commonly referred to as "methane migration". This can be caused by old natural gas wells near water well systems becoming abandoned and no longer monitored.

Lately, however, the described wells/pumps are no longer very efficient and can be replaced by either handpumps or treadle pumps. Another alternative is the use of self-dug wells, electrical deep-well pumps (for higher depths). Appropriate technology organizations as Practical Action are now supplying information on how to build/set-up (DIY) handpumps and treadle pumps in practice.

History

Camel drawing water from a well, Djerba island, Tunisia.

Wood-lined wells are known from the early Neolithic Linear Pottery culture, for example in Kückhoven (an outlying centre of Erkelenz), dated 5090 BC and Eythra, dated 5200 BC in Schletz (an outlying centre of Asparn an der Zaya) in Austria.

Australian Aborigines relied on wells to survive the harsh Australian desert. They would dig down, scooping out sand and mud to reach clean water, then cover the source with spinifex to prevent spoilage. Non-Aborigines call these native wells, soaks or soakages.

Stepwells are common in the west of India. In these wells, the water may be reached by descending a set of steps. They may be covered and are often of architectural significance. Many stepwells were also used for leisure, providing relief from the daytime heat.

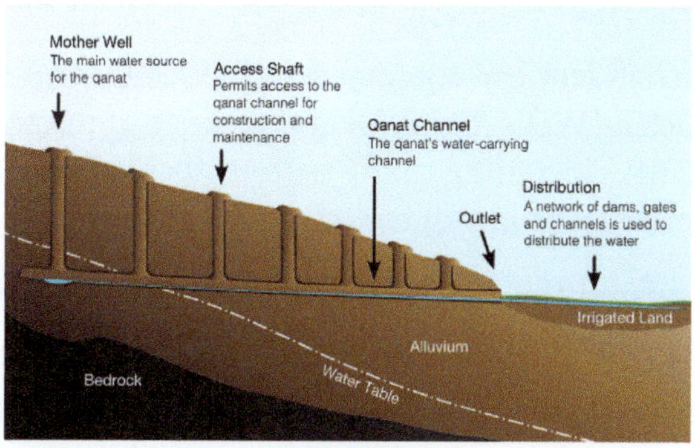

A qanat

A qanat is an ancient water collection system made up of a series of wells and linked underground water channels that collects flowing water from a source usually a distance away, stores it, and then brings the water to the surface using gravity. Much of the population of Iran and other arid countries in Asia and North Africa historically depended upon the water from qanats; the areas of population corresponded closely to the areas where qanats are possible.

In Egypt, shadoofs and sakiehs are used. When compared to each other however, the Sakkieh is much more efficient, as it can bring up water from a depth of 10 metres (versus the 3 metres of the shadoof). The Sakieh is the Egyptian version of the Noria.

From the Iron Age onwards, wells are common archaeological features, both with wooden shafts and shaft linings made from wickerwork.

Some of the world's oldest known wells, located in Cyprus, date to 7000-8500 BC. Two wells from the Neolithic period, around 6500 BC, have been discovered in Israel. One is in Atlit, on the northern coast of Israel, and the other is the Jezreel Valley.

Society and Culture

Water use, tacuinum sanitatis casanatensis (14th century).

Springs and wells have had cultural significance since prehistoric times, leading to the foundation of towns such as Wells and Bath in Somerset. Interest in health benefits led to the growth of spa towns including many with *wells* in their name, examples being Llandrindod Wells and Royal Tunbridge Wells.

Eratosthenes first calculated the radius of the Earth in about 230 BC by comparing shadows in wells during the summer solstice.

Many incidents in the bible take place around wells, such as the finding of a wife for Isaac in Genesis and Jesus's talk with the Samaritan woman in the Gospels.

Rainwater Harvesting

Rainwater capture and storage system at the Monterrey Institute of Technology and Higher Education, Mexico City.

A cistern for rainwater storage

Rainwater harvesting is the accumulation and deposition of rainwater for reuse on-site, rather than allowing it to run off. Rainwater can be collected from rivers or roofs, and in many places the water collected is redirected to a deep pit (well, shaft, or borehole), a reservoir with percolation, or collected from dew or fog with nets or other tools. Its uses include water for gardens, livestock, irrigation, domestic use with proper treatment, and indoor heating for houses etc. The harvested water can also be used as drinking water, longer-term storage and for other purposes such as groundwater recharge.

Rainwater harvesting is one of the simplest and oldest methods of self-supply of water for households usually financed by the user.

Advantages

Rainwater harvesting provides an independent water supply during regional water restrictions and in developed countries is often used to supplement the main supply. It provides water when there is a drought, can help mitigate flooding of low-lying areas, and reduces demand on wells which may enable groundwater levels to be sustained. It also helps in the availability of potable water as rainwater is substantially free of salinity and other salts. Application of rainwater harvesting in urban water system provides a substantial benefit for both water supply and wastewater subsystems by reducing the need for clean water in water distribution system, less generated stormwater in sewer system, as well as a reduction in stormwater runoff polluting freshwater bodies.

There has been a large body of work focused on the development of Life Cycle Assessment and Life Cycle Costing methodologies to assess the level of environmental impacts and money that can be saved by implementing rainwater harvesting systems.

More development and knowledge is required to understand the benefits rainwater harvesting can provide to agriculture. Many countries especially those with an arid environment use rainwater harvesting as a cheap and reliable source of clean water. To enhance irrigation in arid environments, ridges of soil are constructed in order to trap and prevent rainwater from running down hills and slopes. Even in periods of low rainfall, enough water is collected in order for crops to grow. Water can be collected from roofs, dams, and ponds can be constructed in order to hold large quantities of rainwater so that even on days where there is little to no rainfall, there is enough available to irrigate crops.

Quality

The concentration of contaminants is reduced significantly by diverting the initial flow of run-off water to waste. Improved water quality can also be obtained by using a floating draw-off mechanism (rather than from the base of the tank) and by using a series of tanks, withdraw from the last in series. Pre-filtration is a common practice used in the industry to ensure that the water entering the tank is free of large sediment. Pre-filtration is important to keep the system healthy.

Conceptually, a water supply system should match the quality of water with the end use. However, in most of the developed world high quality potable water is used for all end uses. This approach wastes money and energy and imposes unnecessary impacts to the environment. Supplying rainwater that has gone through preliminary filtration measures for non-potable water uses, such as toilet flushing, irrigation, and laundry, may be a significant part of a sustainable water management strategy.

System Setup

Rainwater harvesting systems can range in complexity, from systems that can be installed with minimal skills, to automated systems that require advanced setup and installation. The basic Rainwater harvesting system is more of a plumbing job than a technical job as all the outlets from the building terrace are connected through a pipe to an underground tank that stores water.

Systems are ideally sized to meet the water demand throughout the dry season since it must be big enough to support daily water consumption. Specifically, the rainfall capturing area such as a building roof must be large enough to maintain adequate flow of water. The water storage tank size should be large enough to contain the captured water.

For low-tech systems, there are many low-tech methods used to capture rainwater: rooftop systems, surface water capture, and pumping the rainwater that has already soaked into the ground or captured in reservoirs and storing it into tanks (cisterns).

Before a rainwater harvesting system is built, it is helpful to use digital tools. For instance, if you want to detect if a region has a high rainwater harvesting potential, rainwater harvesting GIS maps can be made using an online interactive tool. Or if you need to estimate how much water is needed to fulfill a community's water needs, the Rain is Gain tool helps with this. Tools like these can save time and money before a commitment to build a system is undertaken, in addition to making the project sustainable and last a long time.

Life Cycle Assessment: Design for Environment

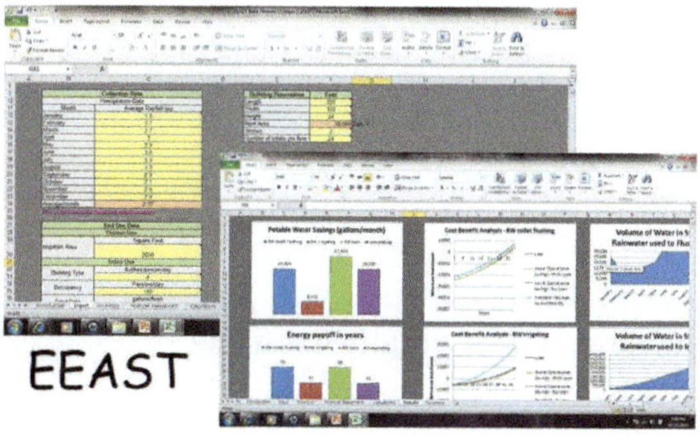

EEAST Model for LCAs of Rainwater Harvesting Systems.

Contemporary system designs require an analysis of not only the economic and technical performance of a system, but also the environmental performance. Life Cycle Assessment is a methodology used to evaluate the environmental impacts of a precut or systems, from cradle-to-grave of its' lifetime. Devkota et al., developed such a methodology for rainwater harvesting, and found that the building design (e.g., dimensions) and function (e.g., educational, residential, etc.) play critical roles in the environmental performance of the system. The Economic and Environmental Analysis of Sanitations Technologies, EEAST model evaluates the greenhouse gas emissions and cost of such systems over the lifetime of a variety of building types.

To address the functional parameters of rainwater harvesting systems, a new metric was developed - the demand to supply ratio (D/S) - identifying the ideal building design (supply) and function (demand) in regard to the environmental performance of rainwater harvesting for toilet flushing. With the idea that supply of rainwater not only saves the potable water but also saves the stormwater entering the combined sewer network (thereby requiring treatment), the savings in environmental emissions were higher if the buildings are connected to a combined sewer network compared to separate one.

Rain Water Harvesting by Freshwater Flooded Forests

Ratagul Freshwater Flooded Forest, Bangladesh

Rainwater harvesting is possible by growing fresh water flooded forests without losing the income from the used /submerged land. The main purpose of the rainwater harvesting is to utilize the locally available rainwater to meet water requirements throughout the year without the need of huge capital expenditure. This would facilitate the availability of uncontaminated water for domestic, industrial and irrigation needs.

New Approaches

Instead of using the roof for catchment, the RainSaucer, which looks like an upside-down umbrella, collects rain straight from the sky. This decreases the potential for contamination and makes potable water for developing countries a potential application. Other applications of this free standing rainwater collection approach are sustainable gardening and small plot farming.

A Dutch invention called the Groasis Waterboxx is also useful for growing trees with harvested and stored dew and rainwater.

Presentation of RainSaucer system to students at Orphanage in Guatemala.

Traditionally, storm water management using detention basins served a single purpose. However, Optimized Real-Time Control (OptiRTC) lets this infrastructure double as a source of rainwater harvesting without compromising the existing detention capacity. This has been used in the EPA headquarters to evacuate stored water prior to storm events, thus reducing wet weather flow while ensuring water availability for later reuse. This has the benefit of increasing water quality released and decreasing the volume of water released during combined sewer overflow events.

Generally, check dams are constructed across the streams to enhance the percolation of surface water into the sub soil strata. The water percolation in the water impounded area of the check dams, can be enhanced artificially many folds by loosening the sub soil strata / overburden by using ANFO explosives as used in open cast mining. Thus local aquifers can be recharged quickly by using the available surface water fully for using in the dry season.

History

Around the third century BC, the farming communities in Balochistan (now located in Pakistan, Afghanistan and Iran), and Kutch, India, used rainwater harvesting for agriculture and many uses also. In ancient Tamil Nadu , rainwater harvesting was done by Chola kings. Rainwater from the Brihadeeswarar temple (located in Balaganpathy Nagar, Thanjavur, India) was collected in Shivaganga tank. During the later Chola period, the Vīrānam tank was built (1011 to 1037 CE) in Cuddalore district of Tamil Nadu state to store water for drinking and irrigation purposes. Vīrānam is a 16 km (9.9 mi) long tank with a storage capacity of 1,465,000,000 cubic feet (41,500,000 m^3).

Rainwater harvesting was done in the Indian states of Madhya Pradesh, Maharashtra, and Chhattisgarh in the olden days. Ratanpur, in the state of Chhattisgarh, had around 150 ponds. Most of the tanks or ponds were utilized in agriculture works.

Rainwater harvesting tank in Burkina Faso

It is a little-known fact that for centuries the town of Venice depended on rainwater harvesting. The lagoon which surrounds Venice is made of brackish water which is not suitable for human drinking. The ancient inhabitants of Venice established a system of rainwater collection which was based on man-made insulated collection wells. Water would percolate down the specially designed stone flooring, and be filtered by a layer of sand, then collecting at the bottom of the well. Later, as Venice acquired territories on the mainland, it started to import water by boat from local rivers, but the wells remained in use, and were especially important in time of war when access to the mainland water could be blocked by an assailant.

Rainwater harvesting system in South Africa

Current Use

- In China, Argentina and Brazil rooftop rainwater harvesting is being practised for providing drinking water, domestic water, water for livestock, water for small irrigation and a way to replenish groundwater levels. Gansu province in China and semi-arid north east Brazil have the largest rooftop rainwater harvesting projects ongoing.
- Thailand is the country where the largest fraction of the population in the rural area rely on rainwater harvesting (currently around 40%). Rainwater harvesting was promoted heavily by the government in the 1980's. In the 1990's, after government funding for the collection tanks ran out, the private sector stepped in and provided several million tanks to private households, many of which continue to be used until now. This is one of the largest example of self-supply of water worldwide.
- In Bermuda, the law requires all new construction to include rainwater harvesting adequate for the residents.
- The U.S. Virgin Islands have a similar law.
- In Senegal and Guinea-Bissau, the houses of the Diola-people are frequently equipped with homebrew rainwater harvesters made from local, organic materials.
- In the Irrawaddy Delta of Myanmar, the groundwater is saline and commu-

nities rely on mud-lined rainwater ponds to meet their drinking water needs throughout the dry season. Some of these ponds are centuries old and are treated with great reverence and respect.

- In the United States: until 2009 in Colorado, water rights laws almost completely restricted rainwater harvesting; a property owner who captured rainwater was deemed to be stealing it from those who have rights to take water from the watershed. Now, residential well owners that meet certain criteria may obtain a permit to install a *rooftop precipitation collection system* (SB 09-080). Up to 10 large scale pilot studies may also be permitted (HB 09-1129). The main factor in persuading the Colorado Legislature to change the law was a 2007 study that found that in an average year, 97% of the precipitation that fell in Douglas County, in the southern suburbs of Denver, never reached a stream—it was used by plants or evaporated on the ground. In Colorado you cannot even drill a water well unless you have at least 35 acres (14 ha). In New Mexico, rainwater catchment is mandatory for new dwellings in Santa Fe. Texas offers a sales tax exemption on the purchase of rainwater harvesting equipment. Both Texas and Ohio allow the practice even for potable purposes. Oklahoma passed the Water for 2060 Act in 2012, to promote pilot projects for rainwater and graywater use among other water saving techniques.
- In Beijing, some housing societies are now adding rainwater in their main water sources after proper treatment.
- In Ireland, Professor Micheal Mcginley established a project to design a rainwater harvesting prototype in the Biosystems design Challenge Module at University College Dublin

Canada

A number of Canadians have started implementing rainwater harvesting systems for use in stormwater reduction, irrigation, laundry, and lavatory plumbing. Substantial reform to Canadian law since the mid-2000s has increased the use of this technology in agricultural, industrial, and residential use; but ambiguity remains amongst legislation in many provinces. Bylaws and local municipal codes often regulate rainwater harvesting.

India

- Pan India Potential:No authenticated potential of rainwater harvesting has been assessed in India. On absolute local basis it has been carried out at some places.
- Tamil Nadu: Tamil Nadu was the first state to make Rainwater Harvesting compulsory for every building to avoid groundwater depletion. The scheme was

launched in 2001 and has been implemented in all rural areas of Tamil Nadu. We can see posters all over Tamil Nadu including Rural areas to create awareness about harvesting Rainwater TN Govt site. It gave excellent results within five years, and slowly every state took it as a role model. Since its implementation, Chennai saw a 50 percent rise in water level in five years and the water quality significantly improved.

- Karnataka: In Bangalore it is mandatory for adoption of rainwater harvesting for every owner or the occupier of a building having the sital area measuring 60 ft (18.3 m) X 40 ft (12.2 m) and above and for newly constructed building measuring 30 ft (9.1 m) X 40 ft (12.2 m) and above dimension. In this regard, BWSSB has initiated and constructed "Rain Water Harvesting Theme Park" in the name of Sir. M. Visvesvaraya in 1.2 acres (4,900 m^2) land situated at Jayanagar, Bangalore. In this park, 26 different type of Rain Water Harvesting models are demonstrated along with the water conservation tips. The Auditorium on the first floor is set up with Green Air conditioning system and the same will be utilized to arrange the meeting and showing the video clip about the rainwater harvesting to students as well as general public.
- Rajasthan: In Rajasthan, rainwater harvesting has traditionally been practised by the people of the Thar Desert. There are many ancient water harvesting systems in Rajasthan, which have now been revived. Water harvesting systems are widely used in other areas of Rajasthan as well, for example the chauka system from the Jaipur district.
- Kerala:
- Maharashtra: At present, in Pune, rainwater harvesting is compulsory for any new housing society to be registered.
- In Mumbai city in Maharashtra rainwater harvesting is being considered as a good solution to solve the water crisis.

The Mumbai city council is planning to make rainwater harvesting mandatory for large societies. An attempt has been made at the Department of Chemical Engineering, IISc, Bangalore to harvest rainwater using upper surface of a solar still, which was used for water distillation

Israel

The Southwest Center for the Study of Hospital and Healthcare Systems in cooperation with Rotary International is sponsoring a rainwater harvesting model program across the country. The first rainwater catchment system was installed at an elementary school in Lod, Israel. The project is looking to expand to Haifa in its third phase. The Southwest Center has also partnered with the Water Resources Action Project (WRAP) of Washington D.C. WRAP currently has rainwater harvesting projects in the West Bank. Rainwater harvesting systems are being installed in local schools for the purpose of educating schoolchildren about water conservation principles and bridging divides between people of different religious and ethnic backgrounds all while addressing the water scarcity issue that the Middle East faces.

New Zealand

Although New Zealand has plentiful rainfall in the West and South, for much of the country, rainwater harvesting is the normal practice for most rural housing and is encouraged by most councils.

Sri Lanka

Rainwater harvesting has been a popular method of obtaining water for agriculture and for drinking purposes in rural homes. The legislation to promote rainwater harvesting was enacted through the Urban Development Authority (Amendment) Act, No. 36 of 2007. Lanka rainwater harvesting forum is leading the Sri Lanka's initiative.

South Africa

The South African Water Research Commission has supported research into rainwater harvesting. Reports on this research are available on their 'Knowledge Hub'. Studies in arid, semi-arid and humid regions have confirmed that techniques such as mulching, pitting, ridging and modified run-on plots are effective for small-scale crop production.

United Kingdom

In the United Kingdom, water butts are often found in domestic gardens and on allotments to collect rainwater, which is then used to water the garden. However, the British government's Code For Sustainable Homes encouraged fitting large underground tanks to new-build homes to collect rainwater for flushing toilets, watering and washing. Ideal designs had the potential to reduce demand on mains water supply by half. The code was revoked in 2015.

United States

- Colorado: Until 2009, water rights laws almost completely restricted rainwater harvesting; a property owner who captured rainwater was deemed to be stealing it from those who have rights to take water from the watershed. Now, residential well owners that meet certain criteria may obtain a permit to install a *rooftop precipitation collection system* (SB 09-080). Up to 10 large scale pilot studies may also be permitted (HB 09-1129). The main factor in persuading the Colorado Legislature to change the law was a 2007 study that found that in an average year, 97% of the precipitation that fell in Douglas County, in the southern suburbs of Denver, never reached a stream—it was used by plants or evaporated on the ground. In Colorado, one is not permitted to drill a water well unless one has at least 35 acres (14 ha) of land.

- New Mexico: In New Mexico, rainwater catchment is mandatory for new dwellings in Santa Fe.
- Texas offers a sales tax exemption on the purchase of rainwater harvesting equipment. Both Texas and Ohio allow the practice even for potable purposes.
- Oklahoma passed the Water for 2060 Act in 2012, to promote pilot projects for rainwater and graywater use among other water saving techniques.

Non-traditional

- In 1992, American artist Michael Jones McKean created an artwork in Omaha, Nebraska at the Bemis Center for Contemporary Art that created a fully sustainable rainbow in the Omaha skyline. The project collected thousands of gallons of rainwater storing the water in six daisy-chained 12,000 gallons tanks. The massive logistical undertaking, during its 5-month span, was one of the largest urban rainwater harvesting sites in the American Mid-West.

Water Transportation

Water transportation is the intentional movement by water over large distances. Methods of transportation fall into three categories:

- Aqueducts, which include pipelines, canals, and tunnels
- container shipment, which includes transport by tank truck, tank car, and tank ship, and
- towing, where a tugboat is used to pull an iceberg or a large water bag along behind it.

Due to its weight, the transportation of water is very energy intensive. Unless it has the assistance of gravity, a canal or long-distance pipeline will need pumping stations at regular intervals. In this regard, the lower friction levels of the canal make it a more economical solution than the pipeline. Water transportation is also very common along rivers and oceans.

Major Water Transportation Projects

The Grand Canal of China, finished in the 7th Century AD and measuring 1,794 kilometres (1,115 mi).

The California Aqueduct, near Sacramento, is 715 km (444 mi) long.

The Great Manmade River is a vast underground network of pipes (1600 kilometres / 994 miles) in the Sahara desert, transporting water from an immense aquifer to the largest cities in the region.

The Keita Integrated Development Project used specially created plows called the Delphino, Treno and Scarabeo to build water catchments. In these catchments, trees were planted which grow on the water flowing through the ditches. The Kimberley Water Source Project is currently underway in Australia to determine the best method of transporting water from the Fitzroy River to the city of Perth. Options being considered include a 3,700 kilometre canal, a pipeline of at least 1,800 kilometres, tankers of 300,000 to 500,000 tonnes, and water bags each carrying between 0.5 and 1.5 gigalitres.

The Goldfields Pipeline built in Western Australia in 1903 was the longest pipeline of its day at 597 kilometres. It supplies water from Perth to Kalgoorlie's gold mining industry.

Manual Water Transportation

Sakka of Mecca, 1779

Historically water was transported by hand in dry countries, by traditional waterers such as the sakkas of Arabia and Bhishti of India.

References

- Philliops, Gregor (11 May 2011). "Erwin's new water tower will be among tallest on East Coast". The Fayetteville Observer. Retrieved 25 February 2012
- Association, American Water Works (2003). Water Sources. American Water Works Association. ISBN 9781583212295
- "Criteria and Guidelines for the "Rainwater Harvesting"" (PDF). Pilot Project Program. Colorado Water Conservation Board (CWCB). January 28, 2010. Retrieved 2012-03-24
- Committee on Environmental Health; Committee on Infectious Diseases (2009). "Drinking water from private wells and risks to children". Pediatrics. 123 (6): 1599–1605. PMID 19482772. doi:10.1542/peds.2009-0751
- Devkota, J.; Schlachter, H.; Anand, C.; Phillips, R.; Apul, Defne (November 2013). "Development and application of EEAST: A lifecycle-based model for use of harvested rainwater and composting toilets in buildings". Journal of Environmental Management. 130: 397–404. doi:10.1016/j.jenvman.2013.09.015
- Rose, Lisa (22 February 2012), "Despite challenge, Union Township water tower remains a Jersey landmark", The Star-Ledger, retrieved 21 February 2012
- Zhu, Qiang; et al. (2015). Rainwater Harvesting for Agriculture and Water Supply. Beijing: Springer. p. 20. ISBN 978-981-287-964-6
- Devkota, Jay; Schlachter, Hannah; Apul, Defne (May 2015). "Life cycle based evaluation of harvested rainwater use in toilets and for irrigation". Journal of Cleaner Production. 95: 311–321. doi:10.1016/j.jclepro.2015.02.021
- Rural Water Supply Network. "Rural Water Supply Network Self-supply site". www.rural-water-supply.net/en/self-supply. Retrieved 2017-03-19
- Meulemans, C. C. E. (1987-09-01). "The Basic Principles of UV–Disinfection of Water". Ozone: Science & Engineering. 9 (4): 299–313. ISSN 0191-9512. doi:10.1080/01919518708552146
- White, John H. Jr. (1993). The American Railroad Freight Car: From the Wood-Car Era to the Coming of Steel. Baltimore: Johns Hopkins University Press. ISBN 0-8018-4404-5. OCLC 26130632
- "Rainwater Collection in Colorado" (PDF). Colorado water law, notices. Colorado Division of Water Resources. Retrieved 2012-03-24

3

Sewage: Treatment and Management

Sewage is a form of waste produced by people. It consists of greywater and blackwater. Sewage if untreated causes widespread water pollution. The process of removing pollutants from sewage water is known as sewage treatment. The chapter strategically encompasses and incorporates the treatment and management of sewage, providing a complete understanding.

Sewage

Sewage (or domestic wastewater or municipal wastewater) is a type of wastewater that is pruduced from a community of people. It is characterized by volume or rate of flow, physical condition, chemical and toxic constituents, and its bacteriologic status (which organisms it contains and in what quantities). It consists mostly of greywater (from sinks, tubs, showers, dishwashers, and clothes washers), blackwater (the water used to flush toilets, combined with the human waste that it flushes away); soaps and detergents; and toilet paper (less so in regions where bidets are widely used instead of paper). Whether it also contains surface runoff depends on the design of sewer system.

Raw sewage arriving at a sewage treatment plant in Syria

All sewage ends up back in the environment (from which its constituents came), by any of several routes. A basic distinction in its route is whether it undergoes sewage treatment to mitigate its effect on the environment before arriving there. Sewage usually travels from a building's plumbing either into a sewer, which will carry it elsewhere, or into an onsite sewage facility (of which there are many kinds). Whether it is combined with surface runoff in the sewer depends on the sewer design (sanitary sewer or combined sewer). Before the 20th century, sewers usually discharged into a body of water such as a stream, river, lake, bay, or ocean. There was no treatment, so the breakdown of the human waste was left to the ecosystem. Today, the goal is that sewers route their contents to a wastewater treatment plant rather than directly to a body of water. In many countries, this is the norm; in many developing countries, it may be a yet-unrealized goal. In general, with passing decades and centuries, humanity seeks to be smarter about the route of sewage on its way back to the environment, in order to reduce environmental degradation and achieve sustainability. Thus other goals of modern sewage routing include handling surface runoff separately from sewage, handling greywater separately from toilet waste, and coping better with abnormal events (such as peaks in use from internal displacement and peaks in stormwater volumes from extreme weather).

Proper collection and safe, nuisance-free disposal of the liquid wastes of a community are legally recognized as a necessity in an urbanized, industrialized society. The reality is, however, that most wastewater produced globally remains untreated causing widespread water pollution, especially in low-income countries: A global estimate by UNDP and UN-Habitat is that 90% of all wastewater generated is released into the environment untreated. In many developing countries the bulk of domestic and industrial wastewater is discharged without any treatment or after primary treatment only.

The term sewage is nowadays regarded as an older term and is being more and more replaced by "wastewater". In general American English usage, the terms "sewage" and "sewerage" mean the same thing. Both words are descended from Old French *assewer*, derived from the Latin *exaquare*, "to drain out (water)". In American technical and professional English usage, "sewerage" refers to the infrastructure that conveys sewage.

Types

Greywater samples before and after treatment in a constructed wetland in Peru.

- The wastewater from residences and institutions, carrying bodily wastes (primarily feces and urine), washing water, food preparation wastes, laundry wastes, and other waste products of normal living, are classed as domestic or sanitary sewage.
- Liquid-carried wastes from stores and service establishments serving the immediate community, termed commercial wastes, are included in the sanitary or domestic sewage category if their characteristics are similar to household flows. Wastes that result from an industrial processes such as the production or manufacture of goods are classed as industrial wastewater, not as sewage.
- Surface runoff, also known as storm flow or overland flow, is that portion of precipitation that runs rapidly over the ground surface to a defined channel. Precipitation absorbs gases and particulates from the atmosphere, dissolves and leaches materials from vegetation and soil, suspends matter from the land, washes spills and debris from urban streets and highways, and carries all these pollutants as wastes in its flow to a collection point.

Pollutants

Organic Pollutants and Nutrients

Sewage is a complex mixture of chemicals, with many distinctive chemical characteristics. These include high concentrations of ammonium, nitrate, phosphorus, high conductivity (due to high dissolved solids), high alkalinity, with pH typically ranging between 7 and 8. The organic matter of sewage is measured by determining its biological oxygen demand (BOD) or the chemical oxygen demand (COD). Sewage also contains the nutrients nitrogen and phosphorus.

Pathogens

Sewage contains human feces, and therefore often contains pathogens of one of the four types:

- bacteria (for example hepatitis A, rotavirus, enteroviruses),
- viruses (for example Salmonella, Shigella, Campylobacter, Vibrio cholera),
- protozoa (for example Entamoeba histolytica, Giardia lamblia, Cryptosporidium parvum) and
- parasites such as helminths and their eggs (e.g. ascaris (roundworm), ancylostoma (hookworm), trichuris (whipworm))

Sewage can be monitored for both disease-causing and benign organisms with a variety of techniques. Traditional techniques involve filtering, staining, and examining samples under a microscope. Much more sensitive and specific testing can be accomplished with DNA sequencing, such as when looking for rare organisms, attempting eradication, testing specifically for drug-resistant strains, or discovering new species. Sequencing DNA from an environmental sample is known as metagenomics.

Micro-pollutants

Sewage also contains environmental persistent pharmaceutical pollutants. Trihalomethanes can also be present as a result of past disinfection.

Sewage has also been analyzed to determine relative rates of use of prescription and illegal drugs among municipal populations.

Health and Environmental Aspects

All categories of sewage are likely to carry pathogenic organisms that can transmit disease to humans and animals. Sewage also contains organic matter that can cause odor and attract flies.

Sewage contains nutrients that may cause eutrophication of receiving water bodies; and can lead to ecotoxicity.

Reuse of Sewage

Increasingly, agriculture is using untreated wastewater for irrigation. Cities provide lucrative markets for fresh produce, so are attractive to farmers. However, because agriculture has to compete for increasingly scarce water resources with industry and municipal users, there is often no alternative for farmers but to use water polluted with urban waste, including sewage, directly to water their crops. There can be significant health hazards related to using water loaded with pathogens in this way, especially if people eat raw vegetables that have been irrigated with the polluted water.

The International Water Management Institute has worked in India, Pakistan, Vietnam, Ghana, Ethiopia, Mexico and other countries on various projects aimed at assessing and reducing risks of wastewater irrigation. They advocate a 'multiple-barrier' approach to wastewater use, where farmers are encouraged to adopt various risk-reducing behaviours. These include ceasing irrigation a few days before harvesting to allow pathogens to die off in the sunlight, applying water carefully so it does not contaminate leaves likely to be eaten raw, cleaning vegetables with disinfectant or allowing fecal sludge used in farming to dry before being used as a human manure. The World Health Organization has developed guidelines for safe water use.

Collection and Disposal

A system of sewer pipes (sewers) collects sewage and takes it for treatment or disposal.

The system of sewers is called *sewerage* or *sewerage system* in British English and *sewage system* in American English. Where a main sewerage system has not been provided, sewage may be collected from homes by pipes into septic tanks or cesspits, where it may be treated or collected in vehicles and taken for treatment or disposal. Properly functioning septic tanks require emptying every 2–5 years depending on the load of the system.

A medieval waste pipe in Stockholm Old Town formerly deposited sewage on the street to be flushed away by rain.

Sewage canal of a medieval house as depicted in 1447 *St. Barbara Altarpiece* in the National Museum in Warsaw.

Sewage and wastewater is also disposed of to rivers, streams, and the sea in many parts of the world. Doing so can lead to serious pollution of the receiving water. This is common in developing countries and may still occur in some developed countries, where septic tank systems are too expensive.

Treatment

Sewage treatment is the process of removing the contaminants from sewage to produce liquid and solid (sludge) suitable for discharge to the environment or for reuse. It is a form of waste management. A septic tank or other on-site wastewater treatment system such as biofilters or constructed wetlands can be used to treat sewage close to where it is created.

Sewage treatment results in sewage sludge which requires sewage sludge treatment before safe disposal or reuse. Under certain circumstances, the treated sewage sludge might be termed "biosolids" and can be used as a fertilizer.

In developed countries sewage collection and treatment is typically subject to local and national regulations and standards.

Legislation

European Union

Council Directive 91/271/EEC on Urban Wastewater Treatment was adopted on 21 May 1991, amended by the Commission Directive 98/15/EC. Commission Decision 93/481/EEC defines the information that Member States should provide the Commission on the state of implementation of the Directive.

Sewerage

Sewerage is the infrastructure that conveys sewage or surface runoff (stormwater, meltwater, rainwater). It encompasses components such as receiving drains, manholes, pumping stations, storm overflows, and screening chambers of the combined sewer or sanitary sewer. Sewerage ends at the entry to a sewage treatment plant or at the point of discharge into the environment. It is the system of pipes, chambers, manholes, etc. that conveys the sewage or storm water.

According to this definition, sewerage and sewage are two different terms. However, at least in American English colloquial usage, both terms are sometimes used interchangeably.

Components and Types

The main part of a sewerage system is made up of large pipes (i.e. the sewers, or "sanitary sewers") that convey the sewage from the point of production to the point of treatment.

Types of sanitary sewer systems which are all usually gravity sewers include:

- Combined sewer
- Simplified sewerage

- Storm drain

Sanitary sewers not relying solely on gravity include:

- Vacuum sewer
- Effluent sewer

Maintenance and Rehabilitation

Severe constraints are applied to sewerage, which may result in premature deterioration. These include root intrusion, joint displacement, cracks and holes formation leading to a significant volume of leakage with an overall risk for the environment and public health. For example, it is estimated that 500 million m^3 of contaminated water per year can leak into soil and ground-water in Germany. The rehabilitation and replacement of damaged sewers is very costly. Annual rehabilitation costs for Los Angeles County are about €400 million, and in Germany, these costs are estimated to be €100 million.

Hydrogen sulfide (H_2S) is indirectly responsible for biogenic sulfide corrosion and consequently, sewerage need rehabilitation works. Various repairs options are available to Owners over a large range of costs and potential durability. One option is the application of a cementitious material based on calcium aluminate cement, after a cleaning of the corroded structure to remove loose material and contaminants in order to expose a sound, rough and clean substrate. Depending of the concrete condition and contamination, the cleaning can range from simple high pressure jet water cleaning (200 bar) up to real hydro-demolition (2000 bars).

One method to ensure sound concrete is exposed is to verify that the surface pH is superior to 10.

As for any concrete repair, the state-of-the-art rules must be followed. After this cleaning step, the cementitious material is applied to the saturated-surface-dry substrate using either:

- Low pressure wet spray: this method is the more common because it does not produce dust and virtually no material is lost by rebound. It utilizes classical facade rotor pump, easily available in the market. The main drawback is the limited pumping distance that cannot exceed 75 meters.
- Spinning head wet spray: this method is similar to the first but the manual spraying is replaced by a spinning head projecting the mortar onto the repaired surface. This method is fast and especially suited for cylindrical chambers like manholes. When a structure is so severely corroded that man entry is a risk, spinning head application permits an "un-manned" consolidation of the manhole.

- High pressure dry spray: this method, also called "shotcrete" or "gunite" is allowing a faster rate of rehabilitation, and also to make a thicker application in a single pass. The main interest of dry shotcrete is the capacity to pump the mortar over a long distance and this is needed when the access points are distant. Longest dry shotcrete distance the authors are aware is a job site in Australia in 2014 where 100% calcium aluminate mortar was air transported over 800 meters before being sprayed. The main drawback with dry shotcrete is the generation of dust and rebound; these could be limited and controlled with appropriate means (pre-moisture ring, adapted aggregate grading, experienced nozzleman, water mist cut-off walls, etc.).

Challenges

Water Table

Sewerage systems often reduce the water table in areas, especially in densely populated areas where rainwater (from house roofs) is directly piped into the sewerage system as opposed to being simply allowed to be absorbed by the soil. In certain areas it has resulted in a significant lowering of the water table. In the example of Belgium, a lowering of the water table by 100 meters has been the result. The freshwater that is accumulated by the sewerage system is then piped to the sea.

In areas where this is a concern, vacuum sewers may be used instead due to the shallow excavation that is possible for them.

Regulations

In many European countries, citizens are obliged to connect their home sanitation to the national sewerage system where possible. This has resulted in large percentages of the population being connected to a sewerage system. For example, the Netherlands have 99% of the population connected to the sewerage system, and 1% has an individual sewage disposal system or treatment system, e.g., septic tank. Others have slightly lower (though still substantial) percentages; e.g., 96% for Germany.

Sewage Treatment

Sewage treatment is the process of removing contaminants from wastewater, primarily from household sewage. It includes physical, chemical, and biological processes to remove these contaminants and produce environmentally safer treated wastewater (or treated effluent). A by-product of sewage treatment is usually a semi-solid waste or slurry, called sewage sludge, that has to undergo further treatment before being suitable for disposal or land application.

Wastewater treatment plant in Massachusetts, United States.

Sewage treatment may also be referred to as wastewater treatment, although the latter is a broader term which can also be applied to purely industrial wastewater. For most cities, the sewer system will also carry a proportion of industrial effluent to the sewage treatment plant which has usually received pretreatment at the factories themselves to reduce the pollutant load. If the sewer system is a combined sewer then it will also carry urban runoff (stormwater) to the sewage treatment plant. Sewage water can travel towards treatment plants via piping and in a flow aided by gravity and pumps. The first part of filtration of sewage typically includes a bar screen to filter solids and large objects which are then collected in dumpsters and disposed of in landfills. Fat and grease will also be removed before the primary treatment of sewage.

Terminology

The term "sewage treatment plant" (or "sewage treatment works" in some countries) is nowadays often replaced with the term "wastewater treatment plant".

Sewage can be treated close to where the sewage is created, which may be called a "decentralized" system or even an "on-site" system (in septic tanks, biofilters or aerobic treatment systems). Alternatively, sewage can be collected and transported by a network of pipes and pump stations to a municipal treatment plant. This is called a "centralized" system.

Origins of Sewage

Sewage is generated by residential, institutional, commercial and industrial establishments. It includes household waste liquid from toilets, baths, showers, kitchens, and sinks draining into sewers. In many areas, sewage also includes liquid waste from industry and commerce. The separation and draining of household waste into greywater and blackwater is becoming more common in the developed world, with treated greywater being permitted to be used for watering plants or recycled for flushing toilets.

Sewage Mixing with Rainwater

Sewage may include stormwater runoff or urban runoff. Sewerage systems capable of handling storm water are known as combined sewer systems. This design was common when urban sewerage systems were first developed, in the late 19th and early 20th centuries. Combined sewers require much larger and more expensive treatment facilities than sanitary sewers. Heavy volumes of storm runoff may overwhelm the sewage treatment system, causing a spill or overflow. Sanitary sewers are typically much smaller than combined sewers, and they are not designed to transport stormwater. Backups of raw sewage can occur if excessive infiltration/inflow (dilution by stormwater and/or groundwater) is allowed into a sanitary sewer system. Communities that have urbanized in the mid-20th century or later generally have built separate systems for sewage (sanitary sewers) and stormwater, because precipitation causes widely varying flows, reducing sewage treatment plant efficiency.

As rainfall travels over roofs and the ground, it may pick up various contaminants including soil particles and other sediment, heavy metals, organic compounds, animal waste, and oil and grease. Some jurisdictions require stormwater to receive some level of treatment before being discharged directly into waterways. Examples of treatment processes used for stormwater include retention basins, wetlands, buried vaults with various kinds of media filters, and vortex separators (to remove coarse solids).

Industrial Effluent

In highly regulated developed countries, industrial effluent usually receives at least pretreatment if not full treatment at the factories themselves to reduce the pollutant load, before discharge to the sewer. This process is called industrial wastewater treatment. The same does not apply to many developing countries where industrial effluent is more likely to enter the sewer if it exists, or even the receiving water body, without pretreatment.

Industrial wastewater may contain pollutants which cannot be removed by conventional sewage treatment. Also, variable flow of industrial waste associated with production cycles may upset the population dynamics of biological treatment units, such as the activated sludge process.

Process Steps

Overview

Sewage collection and treatment is typically subject to local, state and federal regulations and standards.

Treating wastewater has the aim to produce an effluent that will do as little harm as possible when discharged to the surrounding environment, thereby preventing pollution compared to releasing untreated wastewater into the environment.

Sewage treatment generally involves three stages, called primary, secondary and tertiary treatment.

- *Primary treatment* consists of temporarily holding the sewage in a quiescent basin where heavy solids can settle to the bottom while oil, grease and lighter solids float to the surface. The settled and floating materials are removed and the remaining liquid may be discharged or subjected to secondary treatment. Some sewage treatment plants that are connected to a combined sewer system have a bypass arrangement after the primary treatment unit. This means that during very heavy rainfall events, the secondary and tertiary treatment systems can be bypassed to protect them from hydraulic overloading, and the mixture of sewage and stormwater only receives primary treatment.

- *Secondary treatment* removes dissolved and suspended biological matter. Secondary treatment is typically performed by indigenous, water-borne micro-organisms in a managed habitat. Secondary treatment may require a separation process to remove the micro-organisms from the treated water prior to discharge or tertiary treatment.

- *Tertiary treatment* is sometimes defined as anything more than primary and secondary treatment in order to allow rejection into a highly sensitive or fragile ecosystem (estuaries, low-flow rivers, coral reefs,...). Treated water is sometimes disinfected chemically or physically (for example, by lagoons and microfiltration) prior to discharge into a stream, river, bay, lagoon or wetland, or it can be used for the irrigation of a golf course, green way or park. If it is sufficiently clean, it can also be used for groundwater recharge or agricultural purposes.

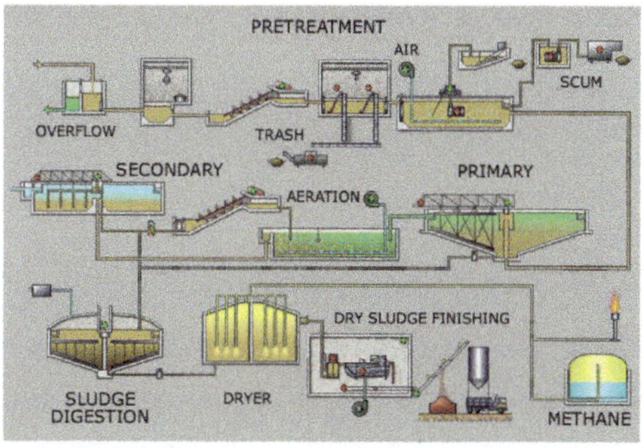

Simplified process flow diagram for a typical large-scale treatment plant.

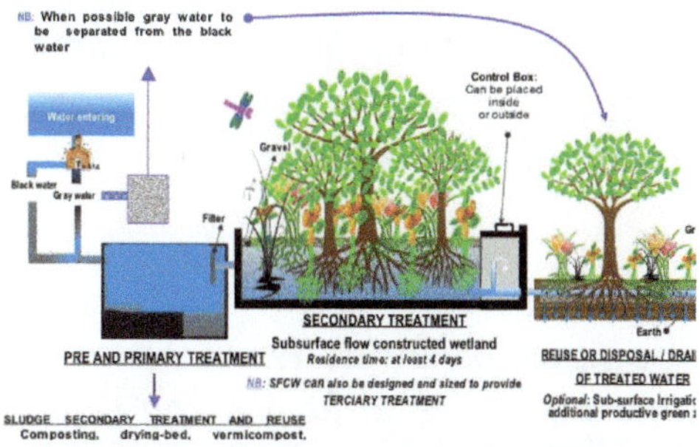

Process flow diagram for a typical treatment plant via subsurface flow constructed wetlands (SFCW).

Pretreatment

Pretreatment removes all materials that can be easily collected from the raw sewage before they damage or clog the pumps and sewage lines of primary treatment clarifiers. Objects commonly removed during pretreatment include trash, tree limbs, leaves, branches, and other large objects.

The influent in sewage water passes through a bar screen to remove all large objects like cans, rags, sticks, plastic packets etc. carried in the sewage stream. This is most commonly done with an automated mechanically raked bar screen in modern plants serving large populations, while in smaller or less modern plants, a manually cleaned screen may be used. The raking action of a mechanical bar screen is typically paced according to the accumulation on the bar screens and/or flow rate. The solids are collected and later disposed in a landfill, or incinerated. Bar screens or mesh screens of varying sizes may be used to optimize solids removal. If gross solids are not removed, they become entrained in pipes and moving parts of the treatment plant, and can cause substantial damage and inefficiency in the process.

Grit Removal

Pretreatment may include a sand or grit channel or chamber, where the velocity of the incoming sewage is adjusted to allow the settlement of sand, grit, stones, and broken glass. These particles are removed because they may damage pumps and other equipment. For small sanitary sewer systems, the grit chambers may not be necessary, but grit removal is desirable at larger plants. Grit chambers come in 3 types: horizontal grit chambers, aerated grit chambers and vortex grit chambers. The process is called sedimentation.

Flow Equalization

Clarifiers and mechanized secondary treatment are more efficient under uniform flow conditions. Equalization basins may be used for temporary storage of diurnal or wet-weather flow peaks. Basins provide a place to temporarily hold incoming sewage during plant maintenance and a means of diluting and distributing batch discharges of toxic or high-strength waste which might otherwise inhibit biological secondary treatment (including portable toilet waste, vehicle holding tanks, and septic tank pumpers). Flow equalization basins require variable discharge control, typically include provisions for bypass and cleaning, and may also include aerators. Cleaning may be easier if the basin is downstream of screening and grit removal.

Fat and Grease Removal

In some larger plants, fat and grease are removed by passing the sewage through a small tank where skimmers collect the fat floating on the surface. Air blowers in the base of the tank may also be used to help recover the fat as a froth. Many plants, however, use primary clarifiers with mechanical surface skimmers for fat and grease removal.

Primary Treatment

Primary treatment tanks in Oregon, USA

In the primary sedimentation stage, sewage flows through large tanks, commonly called "pre-settling basins", "primary sedimentation tanks" or "primary clarifiers". The tanks are used to settle sludge while grease and oils rise to the surface and are skimmed off. Primary settling tanks are usually equipped with mechanically driven scrapers that continually drive the collected sludge towards a hopper in the base of the tank where it is pumped to sludge treatment facilities. Grease and oil from the floating material can sometimes be recovered for saponification (soap making).

Secondary Treatment

Secondary treatment is designed to substantially degrade the biological content of the sewage which are derived from human waste, food waste, soaps and detergent. The majority of municipal plants treat the settled sewage liquor using aerobic biological processes. To be effective, the biota require both oxygen and food to live. The bacteria and protozoa consume biodegradable soluble organic contaminants (e.g. sugars, fats, organic short-chain carbon molecules, etc.) and bind much of the less soluble fractions into floc.

Secondary treatment systems are classified as fixed-film or suspended-growth systems.

- Fixed-film or attached growth systems include trickling filters, constructed wetlands, bio-towers, and rotating biological contactors, where the biomass grows on media and the sewage passes over its surface. The fixed-film principle has further developed into Moving Bed Biofilm Reactors (MBBR) and Integrated Fixed-Film Activated Sludge (IFAS) processes. An MBBR system typically requires a smaller footprint than suspended-growth systems.
- Suspended-growth systems include activated sludge, where the biomass is mixed with the sewage and can be operated in a smaller space than trickling filters that treat the same amount of water. However, fixed-film systems are more able to cope with drastic changes in the amount of biological material and can provide higher removal rates for organic material and suspended solids than suspended growth systems.

Secondary clarifier at a rural treatment plant

Some secondary treatment methods include a secondary clarifier to settle out and separate biological floc or filter material grown in the secondary treatment bioreactor.

Tertiary Treatment

The purpose of tertiary treatment is to provide a final treatment stage to further improve the effluent quality before it is discharged to the receiving environment (sea, river, lake, wet lands, ground, etc.). More than one tertiary treatment process may be used at any treatment plant. If disinfection is practised, it is always the final process. It is also called "effluent polishing."

Filtration

Sand filtration removes much of the residual suspended matter. Filtration over activated carbon, also called *carbon adsorption,* removes residual toxins.

Lagoons or Ponds

A sewage treatment plant and lagoon in Everett, Washington, United States.

Lagoons or ponds provide settlement and further biological improvement through storage in large man-made ponds or lagoons. These lagoons are highly aerobic and colonization by native macrophytes, especially reeds, is often encouraged. Small filter-feeding invertebrates such as *Daphnia* and species of *Rotifera* greatly assist in treatment by removing fine particulates.

Biological Nutrient Removal

Biological nutrient removal (BNR) is regarded by some as a type of secondary treatment process, and by others as a tertiary (or "advanced") treatment process.

Wastewater may contain high levels of the nutrients nitrogen and phosphorus. Excessive release to the environment can lead to a buildup of nutrients, called eutrophication, which can in turn encourage the overgrowth of weeds, algae, and cyanobacteria (blue-green algae). This may cause an algal bloom, a rapid growth in the population of algae. The algae numbers are unsustainable and eventually most of them die. The decomposition of the algae by bacteria uses up so much of the oxygen in the water that most or all of the animals die, which creates more organic matter for the bacteria to decompose. In addition to causing deoxygenation, some algal species produce toxins that contaminate drinking water supplies. Different treatment processes are required to remove nitrogen and phosphorus.

Nitrogen Removal

Nitrogen is removed through the biological oxidation of nitrogen from ammonia to nitrate (nitrification), followed by denitrification, the reduction of nitrate to nitrogen gas. Nitrogen gas is released to the atmosphere and thus removed from the water.

Nitrification itself is a two-step aerobic process, each step facilitated by a different type of bacteria. The oxidation of ammonia (NH_3) to nitrite (NO_2^-) is most often facilitated by *Nitrosomonas* spp. ("nitroso" referring to the formation of a nitroso functional group). Nitrite oxidation to nitrate (NO_3^-), though traditionally believed to be facilitated by *Nitrobacter* spp. (nitro referring the formation of a nitro functional group), is now known to be facilitated in the environment almost exclusively by *Nitrospira* spp.

Denitrification requires anoxic conditions to encourage the appropriate biological communities to form. It is facilitated by a wide diversity of bacteria. Sand filters, lagooning and reed beds can all be used to reduce nitrogen, but the activated sludge process (if designed well) can do the job the most easily. Since denitrification is the reduction of nitrate to dinitrogen (molecular nitrogen) gas, an electron donor is needed. This can be, depending on the waste water, organic matter (from feces), sulfide, or an added donor like methanol. The sludge in the anoxic tanks (denitrification tanks) must be mixed well (mixture of recirculated mixed liquor, return activated sludge [RAS], and raw influent) e.g. by using submersible mixers in order to achieve the desired denitrification.

Sometimes the conversion of toxic ammonia to nitrate alone is referred to as tertiary treatment.

Over time, different treatment configurations have evolved as denitrification has become more sophisticated. An initial scheme, the Ludzack-Ettinger Process, placed an anoxic treatment zone before the aeration tank and clarifier, using the return activated sludge (RAS) from the clarifier as a nitrate source. Influent wastewater (either raw or as effluent from primary clarification) serves as the electron source for the facultative bacteria to metabolize carbon, using the inorganic nitrate as a source of oxygen instead of dissolved molecular oxygen. This denitrification scheme was naturally limited to the amount of soluble nitrate present in the RAS. Nitrate reduction was limited because RAS rate is limited by the performance of the clarifier.

The "Modified Ludzak-Ettinger Process" (MLE) is an improvement on the original concept, for it recycles mixed liquor from the discharge end of the aeration tank to the head of the anoxic tank to provide a consistent source of soluble nitrate for the facultative bacteria. In this instance, raw wastewater continues to provide the electron source, and sub-surface mixing maintains the bacteria in contact with both electron source and soluble nitrate in the absence of dissolved oxygen.

Many sewage treatment plants use centrifugal pumps to transfer the nitrified mixed liquor from the aeration zone to the anoxic zone for denitrification. These pumps are often referred to as *Internal Mixed Liquor Recycle* (IMLR) pumps. IMLR may be 200% to 400% the flow rate of influent wastewater (Q.) This is in addition to Return Activated Sludge (RAS) from secondary clarifiers, which may be 100% of Q. (Therefore, the hydraulic capacity of the tanks in such a system should handle at least 400% of annual average design flow (AADF.) At times, the raw or primary effluent wastewater must be carbon-supplemented by the addition of methanol, acetate, or simple food waste (molasses, whey, plant starch) to improve the treatment efficiency. These carbon additions should be accounted for in the design of a treatment facility's organic loading.

Further modifications to the MLE were to come: Bardenpho and Biodenipho processes include additional anoxic and oxidative processes to further polish the conversion of nitrate ion to molecular nitrogen gas. Use of an anaerobic tank following the initial anoxic process allows for luxury uptake of phosphorus by bacteria, thereby biologically reducing orthophosphate ion in the treated wastewater. Even newer improvements, such as Anammox Process, interrupt the formation of nitrate at the nitrite stage of nitrification, shunting nitrite-rich mixed liquor activated sludge to treatment where nitrite is then converted to molecular nitrogen gas, saving energy, alkalinity, and secondary carbon sourcing. Anammox™ (ANaerobic AMMonia OXidation) works by artificially extending detention time and preserving denitrifiying bacteria through the use of substrate added to the mixed liquor and continuously recycled from it prior to secondary clarification. Many other proprietary schemes are being deployed, including DEMON™, Sharon-ANAMMOX™, ANITA-Mox™, and DeAmmon™. The bacteria Brocadia anammoxidans can remove ammonium from waste water through anaerobic oxidation of ammonium to hydrazine, a form of rocket fuel.

Phosphorus Removal

Every adult human excretes between 200 and 1000 grams of phosphorus annually.

Studies of United States sewage in the late 1960s estimated mean per capita contributions of 500 grams in urine and feces, 1000 grams in synthetic detergents, and lesser variable amounts used as corrosion and scale control chemicals in water supplies. Source control via alternative detergent formulations has subsequently reduced the largest contribution, but the content of urine and feces will remain unchanged. Phosphorus removal is important as it is a limiting nutrient for algae growth in many fresh water systems. It is also particularly important for water reuse systems where high phosphorus concentrations may lead to fouling of downstream equipment such as reverse osmosis.

Phosphorus can be removed biologically in a process called enhanced biological phosphorus removal. In this process, specific bacteria, called polyphosphate-accumulating organisms (PAOs), are selectively enriched and accumulate large quantities of phosphorus within their cells (up to 20 percent of their mass). When the biomass enriched in these bacteria is separated from the treated water, these biosolids have a high fertilizer value.

Phosphorus removal can also be achieved by chemical precipitation, usually with salts of iron (e.g. ferric chloride), aluminum (e.g. alum), or lime. This may lead to excessive sludge production as hydroxides precipitates and the added chemicals can be expensive. Chemical phosphorus removal requires significantly smaller equipment footprint than biological removal, is easier to operate and is often more reliable than biological phosphorus removal. Another method for phosphorus removal is to use granular laterite.

Once removed, phosphorus, in the form of a phosphate-rich sewage sludge, may be dumped in a landfill or used as fertilizer. In the latter case, the treated sewage sludge is also sometimes referred to as biosolids.

Disinfection

The purpose of disinfection in the treatment of waste water is to substantially reduce the number of microorganisms in the water to be discharged back into the environment for the later use of drinking, bathing, irrigation, etc. The effectiveness of disinfection depends on the quality of the water being treated (e.g., cloudiness, pH, etc.), the type of disinfection being used, the disinfectant dosage (concentration and time), and other environmental variables. Cloudy water will be treated less successfully, since solid matter can shield organisms, especially from ultraviolet light or if contact times are low. Generally, short contact times, low doses and high flows all militate against effective disinfection. Common methods of disinfection include ozone, chlorine, ultraviolet light, or sodium hypochlorite. Chloramine, which is used for drinking water, is not used in the treatment of waste water because of its persistence. After multiple steps of disinfection, the treated water is ready to be released back into the water cycle by means of the nearest body of water or agriculture. Afterwards, the water can be transferred to reserves for everyday human uses.

Chlorination remains the most common form of waste water disinfection in North America due to its low cost and long-term history of effectiveness. One disadvantage is that chlorination of residual organic material can generate chlorinated-organic compounds that may be carcinogenic or harmful to the environment. Residual chlorine or chloramines may also be capable of chlorinating organic material in the natural aquatic environment. Further, because residual chlorine is toxic to aquatic species, the treated effluent must also be chemically dechlorinated, adding to the complexity and cost of treatment.

Ultraviolet (UV) light can be used instead of chlorine, iodine, or other chemicals. Because no chemicals are used, the treated water has no adverse effect on organisms that later consume it, as may be the case with other methods. UV radiation causes damage to the genetic structure of bacteria, viruses, and other pathogens, making them incapable of reproduction. The key disadvantages of UV disinfection are the need for frequent lamp maintenance and replacement and the need for a highly treated effluent to ensure that the target microorganisms are not shielded from the UV radiation (i.e., any solids present in the treated effluent may protect microorganisms from the UV light). In the United Kingdom, UV light is becoming the most common means of disinfection because of the concerns about the impacts of chlorine in chlorinating residual organics in the wastewater and in chlorinating organics in the receiving water. Some sewage treatment systems in Canada and the US also use UV light for their effluent water disinfection.

Ozone (O_3) is generated by passing oxygen (O_2) through a high voltage potential resulting in a third oxygen atom becoming attached and forming O_3. Ozone is very unstable and reactive and oxidizes most organic material it comes in contact with, thereby destroying many pathogenic microorganisms. Ozone is considered to be safer than chlorine because, unlike chlorine which has to be stored on site (highly poisonous in the event of an accidental release), ozone is generated on-site as needed. Ozonation also produces fewer disinfection by-products than chlorination. A disadvantage of ozone disinfection is the high cost of the ozone generation equipment and the requirements for special operators.

Fourth Treatment Stage

Micropollutants such as pharmaceuticals, ingredients of household chemicals, chemicals used in small businesses or industries, environmental persistent pharmaceutical pollutant (EPPP) or pesticides may not be eliminated in the conventional treatment process (primary, secondary and tertiary treatment) and therefore lead to water pollution. Although concentrations of those substances and their decompostion products are quite low, there is still a chance to harm aquatic organisms. For pharmaceuticals, the following substances have been identified as "toxicologically relevant": substances with endocrine disrupting effects, genotoxic substances and substances that enhance the development of bacterial resistances. They mainly belong to the group of environmental persistent pharmaceutical pollutants. Techniques for elimination of micropollutants via a fourth treatment stage during sewage treatment are being tested in Germany, Switzerland and the Netherlands. However, since those techniques are still costly, they are not yet applied on a regular basis. Such process steps mainly consist of activated carbon filters that adsorb the micropollutants. Ozone can also be applied as an oxidative method. Also the use of enzymes such as the enzyme laccase is under investigation. A new concept which could provide an energy-efficient treatment of micropollutants could be the use of laccase secreting fungi cultivated at a wastewater treatment plant to degrade micropollutants and at the same time to provide enzymes at a cathode of a microbial biofuel cells. Microbial biofuel cells are investigated for their property to treat organic matter in wastewater.

To reduce pharmaceuticals in water bodies, also "source control" measures are under investigation, such as innovations in drug development or more responsible handling of drugs.

Odor Control

Odors emitted by sewage treatment are typically an indication of an anaerobic or "septic" condition. Early stages of processing will tend to produce foul-smelling gases, with hydrogen sulfide being most common in generating complaints. Large process plants in urban areas will often treat the odors with carbon reactors, a contact media with bio-slimes, small doses of chlorine, or circulating fluids to biologically capture and metabolize the noxious gases. Other methods of odor control exist, including addition of iron salts, hydrogen peroxide, calcium nitrate, etc. to manage hydrogen sulfide levels.

High-density solids pumps are suitable for reducing odors by conveying sludge through hermetic closed pipework.

Energy Requirements

For conventional sewage treatment plants, around 30 percent of the annual operating costs is usually required for energy. The energy requirements vary with type of treatment process as well as wastewater load. For example, constructed wetlands have a lower energy requirement than activated sludge plants, as less energy is required for the aeration step. Sewage treatment plants that produce biogas in their sewage sludge treatment process with anaerobic digestion can produce enough energy to meet most of the energy needs of the sewage treatment plant itself.

In conventional secondary treatment processes, most of the electricity is used for aeration, pumping systems and equipment for the dewatering and drying of sewage sludge. Advanced wastewater treatment plants, e.g. for nutrient removal, require more energy than plants that only achieve primary or secondary treatment.

Sludge Treatment and Disposal

The sludges accumulated in a wastewater treatment process must be treated and disposed of in a safe and effective manner. The purpose of digestion is to reduce the amount of organic matter and the number of disease-causing microorganisms present in the solids. The most common treatment options include anaerobic digestion, aerobic digestion, and composting. Incineration is also used, albeit to a much lesser degree.

Sludge treatment depends on the amount of solids generated and other site-specific conditions. Composting is most often applied to small-scale plants with aerobic digestion for mid-sized operations, and anaerobic digestion for the larger-scale operations.

The sludge is sometimes passed through a so-called pre-thickener which de-waters the sludge. Types of pre-thickeners include centrifugal sludge thickeners rotary drum sludge thickeners and belt filter presses. Dewatered sludge may be incinerated or transported offsite for disposal in a landfill or use as an agricultural soil amendment.

Environment Aspects

The outlet of the Karlsruhe sewage treatment plant flows into the Alb.

Many processes in a wastewater treatment plant are designed to mimic the natural treatment processes that occur in the environment, whether that environment is a natural water body or the ground. If not overloaded, bacteria in the environment will consume organic contaminants, although this will reduce the levels of oxygen in the water and may significantly change the overall ecology of the receiving water. Native bacterial populations feed on the organic contaminants, and the numbers of disease-causing microorganisms are reduced by natural environmental conditions such as predation or exposure to ultraviolet radiation. Consequently, in cases where the receiving environment provides a high level of dilution, a high degree of wastewater treatment may not be required. However, recent evidence has demonstrated that very low levels of specific contaminants in wastewater, including hormones (from animal husbandry and residue from human hormonal contraception methods) and synthetic materials such as phthalates that mimic hormones in their action, can have an unpredictable adverse impact on the natural biota and potentially on humans if the water is re-used for drinking water. In the US and EU, uncontrolled discharges of wastewater to the environment are not permitted under law, and strict water quality requirements are to be met, as clean drinking water is essential. A significant threat in the coming decades will be the increasing uncontrolled discharges of wastewater within rapidly developing countries.

Effects on Biology

Sewage treatment plants can have multiple effects on nutrient levels in the water that the treated sewage flows into. These nutrients can have large effects on the biological life in the water in contact with the effluent. Stabilization ponds (or sewage treatment ponds) can include any of the following:

- Oxidation ponds, which are aerobic bodies of water usually 1–2 meters in depth that receive effluent from sedimentation tanks or other forms of primary treatment.
    - Dominated by algae
- Polishing ponds are similar to oxidation ponds but receive effluent from an oxidation pond or from a plant with an extended mechanical treatment.
    - Dominated by zooplankton
- Facultative lagoons, raw sewage lagoons, or sewage lagoons are ponds where sewage is added with no primary treatment other than coarse screening. These ponds provide effective treatment when the surface remains aerobic; although anaerobic conditions may develop near the layer of settled sludge on the bottom of the pond.
- Anaerobic lagoons are heavily loaded ponds.
    - Dominated by bacteria
- Sludge lagoons are aerobic ponds, usually 2 to 5 meters in depth, that receive anaerobically digested primary sludge, or activated secondary sludge under water.
    - Upper layers are dominated by algae

Phosphorus limitation is a possible result from sewage treatment and results in flagellate-dominated plankton, particularly in summer and fall.

A phytoplankton study found high nutrient concentrations linked to sewage effluents. High nutrient concentration leads to high chlorophyll a concentrations, which is a proxy for primary production in marine environments. High primary production means high phytoplankton populations and most likely high zooplankton populations, because zooplankton feed on phytoplankton. However, effluent released into marine systems also leads to greater population instability.

The planktonic trends of high populations close to input of treated sewage is contrasted by the bacterial trend. In a study of *Aeromonas* spp. in increasing distance from a wastewater source, greater change in seasonal cycles was found the furthest from the effluent. This trend is so strong that the furthest location studied actually had an inversion of the *Aeromonas* spp. cycle in comparison to that of fecal coliforms. Since there is a main pattern in the cycles that occurred simultaneously at all stations it indicates seasonal factors (temperature, solar radiation, phytoplankton) control of the bacterial population. The effluent dominant species changes from *Aeromonas caviae* in winter to *Aeromonas sobria* in the spring and fall while the inflow dominant species is *Aeromonas caviae*, which is constant throughout the seasons.

Reuse

With suitable technology, it is possible to reuse sewage effluent for drinking water, although this is usually only done in places with limited water supplies, such as Windhoek and Singapore.

In arid countries, treated wastewater is often used in agriculture. For example in Israel, about 50 percent of agricultural water use (total use was 1 billion cubic metres in 2008) is provided through reclaimed sewer water. Future plans call for increased use of treated sewer water as well as more desalination plants as part of water supply and sanitation in Israel.

Constructed wetlands fed by wastewater provide both treatment and habitats for flora and fauna. Another example for reuse combined with treatment of sewage are the East Kolkata Wetlands in India. These wetlands are used to treat Kolkata's sewage, and the nutrients contained in the wastewater sustain fish farms and agriculture.

Developing Countries

Few reliable figures exist on the share of the wastewater collected in sewers that is being treated in the world. A global estimate by UNDP and UN-Habitat is that 90% of all wastewater generated is released into the environment untreated. In many developing countries the bulk of domestic and industrial wastewater is discharged without any treatment or after primary treatment only.

In Latin America about 15 percent of collected wastewater passes through treatment plants (with varying levels of actual treatment). In Venezuela, a below average country in South America with respect to wastewater treatment, 97 percent of the country's sewage is discharged raw into the environment. In Iran, a relatively developed Middle Eastern country, the majority of Tehran's population has totally untreated sewage injected to the city's groundwater. However, the construction of major parts of the sewage system, collection and treatment, in Tehran is almost complete, and under development, due to be fully completed by the end of 2012. In Isfahan, Iran's third largest city, sewage treatment was started more than 100 years ago.

Only few cities in sub-Saharan Africa have sewer-based sanitation systems, let alone wastewater treatment plants, an exception being South Africa and – until the late 1990s- Zimbabwe. Instead, most urban residents in sub-Saharan Africa rely on on-site sanitation systems without sewers, such as septic tanks and pit latrines, and fecal sludge management in these cities is an enormous challenge.

History

The Great Stink of 1858 stimulated research into the problem of sewage treatment. In this caricature in *The Times*, Michael Faraday reports to *Father Thames* on the state of the river.

Basic sewer systems were used for waste removal in ancient Mesopotamia, where vertical shafts carried the waste away into cesspools. Similar systems existed in the Indus Valley civilization in modern-day India and in Ancient Crete and Greece. In the Middle Ages the sewer systems built by the Romans fell into disuse and waste was collected into cesspools that were periodically emptied by workers known as 'rakers' who would often sell it as fertilizer to farmers outside the city.

Modern sewage systems were first built in the mid-nineteenth century as a reaction to the exacerbation of sanitary conditions brought on by heavy industrialization and urbanization. Due to the contaminated water supply, cholera outbreaks occurred in 1832, 1849 and 1855 in London, killing tens of thousands of people. This, combined with the Great Stink of 1858, when the smell of untreated human waste in the River Thames became overpowering, and the report into sanitation reform of the Royal Commissioner Edwin Chadwick, led to the Metropolitan Commission of Sewers appointing Sir Joseph Bazalgette to construct a vast underground sewage system for the safe removal of waste. Contrary to Chadwick's recommendations, Bazalgette's system, and others later built in Continental Europe, did not pump the sewage onto farm land for use as fertilizer; it was simply piped to a natural waterway away from population centres, and pumped back into the environment.

Early Attempts

One of the first attempts at diverting sewage for use as a fertilizer in the farm was made by the cotton mill owner James Smith in the 1840s. He experimented with a piped distribution system initially proposed by James Vetch that collected sewage from his factory and pumped it into the outlying farms, and his success was enthusiastically followed by Edwin Chadwick and supported by organic chemist Justus von Liebig.

The idea was officially adopted by the Health of Towns Commission, and various schemes (known as sewage farms) were trialled by different municipalities over the next 50 years. At first, the heavier solids were channeled into ditches on the side of the farm and were covered over when full, but soon flat-bottomed tanks were employed as reservoirs for the sewage; the earliest patent was taken out by William Higgs in 1846 for "tanks or reservoirs in which the contents of sewers and drains from cities, towns and villages are to be collected and the solid animal or vegetable matters therein contained, solidified and dried..." Improvements to the design of the tanks included the introduction of the horizontal-flow tank in the 1850s and the radial-flow tank in 1905. These tanks had to be manually de-sludged periodically, until the introduction of automatic mechanical de-sludgers in the early 1900s.

The precursor to the modern septic tank was the cesspool in which the water was sealed off to prevent contamination and the solid waste was slowly liquified due to anaerobic action; it was invented by L.H Mouras in France in the 1860s. Donald Cameron, as City Surveyor for Exeter patented an improved version in 1895, which he called a 'septic tank'; septic having the meaning of 'bacterial'. These are still in worldwide use, especially in rural areas unconnected to large-scale sewage systems.

Chemical Treatment

Sir Edward Frankland, a distinguished chemist,
who demonstrated the possibility of chemically treating sewage in the 1870s.

It was not until the late 19th century that it became possible to treat the sewage by chemically breaking it down through the use of microorganisms and removing the pollutants. Land treatment was also steadily becoming less feasible, as cities grew and the volume of sewage produced could no longer be absorbed by the farmland on the outskirts.

Sir Edward Frankland conducted experiments at the Sewage Farm in Croydon, England, during the 1870s and was able to demonstrate that filtration of sewage through porous gravel produced a nitrified effluent (the ammonia was converted into nitrate) and that the filter remained unclogged over long periods of time. This established the then revolutionary possibility of biological treatment of sewage using a contact bed to oxidize the waste. This concept was taken up by the chief chemist for the London Metropolitan Board of Works, William Libdin, in 1887:

> in all probability the true way of purifying sewage will be first to separate the sludge, and then turn into neutral effluent retain it for a sufficient period, during which time it should be fully aerated, and finally discharge it into the stream in a purified condition. This is indeed what is aimed at and imperfectly accomplished on a sewage farm.

From 1885 to 1891 filters working on this principle were constructed throughout the UK and the idea was also taken up in the US at the Lawrence Experiment Station in Massachusetts, where Frankland's work was confirmed. In 1890 the LES developed a 'trickling filter' that gave a much more reliable performance.

Contact beds were developed in Salford, Lancashire and by scientists working for the London City Council in the early 1890s. According to Christopher Hamlin, this was part of a conceptual revolution that replaced the philosophy that saw "sewage purification as the prevention of decomposition with one that tried to facilitate the biological process that destroy sewage naturally."

Contact beds were tanks containing the inert substance, such as stones or slate, that maximized the surface area available for the microbial growth to break down the sewage. The sewage was held in the tank until it was fully decomposed and it was then filtered out into the ground. This method quickly became widespread, especially in the UK, where it was used in Leicester, Sheffield, Manchester and Leeds. The bacterial bed was simultaneously developed by Joseph Corbett as Borough Engineer in Salford and experiments in 1905 showed that his method was superior in that greater volumes of sewage could be purified better for longer periods of time than could be achieved by the contact bed.

The Royal Commission on Sewage Disposal published its eighth report in 1912 that set what became the international standard for sewage discharge into rivers; the '20:30 standard', which allowed 20 mg Biochemical oxygen demand and 30 mg suspended solid per litre.

References

- Wastewater engineering : treatment and reuse (4th ed.). Metcalf & Eddy, Inc., McGraw Hill, USA. 2003. p. 1807. ISBN 0-07-112250-8
- Environment Agency (archive) – Persistent, bioaccumulative and toxic PBT substances at the Wayback Machine (archived August 4, 2006). environment-agency.gov.uk. Retrieved on 2012-12-19

- Logan, B.; Regan, J. (2006). "Microbial Fuel Cells—Challenges and Applications". Environmental Science & Technology. 40 (17): 5172–5180. doi:10.1021/es0627592
- World Health Organization (2006). Guidelines for the safe use of wastewater, excreta, and greywater. World Health Organization. p. 31. ISBN 9241546859. OCLC 71253096
- Haughey, A. (1968). "The Planktonic Algae of Auckland Sewage Treatment Ponds". New Zealand Journal of Marine and Freshwater Research. 2 (4): 721–766. doi:10.1080/00288330.1968.9515271
- Natural Environmental Research Council – River sewage pollution found to be disrupting fish hormones. Planetearth.nerc.ac.uk. Retrieved on 2012-12-19
- Lienert, J.; Bürki, T.; Escher, B.I. (2007). "Reducing micropollutants with source control: Substance flow analysis of 212 pharmaceuticals in faeces and urine". Water Science & Technology. 56 (5): 87–96. doi:10.2166/wst.2007.560
- Martin V. Melosi (2010). The Sanitary City: Environmental Services in Urban America from Colonial Times to the Present. University of Pittsburgh Press. p. 110. ISBN 9780822973379
- Caribbean Environment Programme (1998). Appropriate Technology for Sewage Pollution Control in the Wider Caribbean Region (PDF). Kingston, Jamaica: United Nations Environment Programme. Retrieved 2009-10-12
- Caperon, J.; Cattell, S. A. & Krasnick, G. (1971). "Phytoplankton Kinetics in a Subtropical Estuary: Eutrophication" (PDF). Limnology and Oceanography. 16 (4): 599–607. doi:10.4319/lo.1971.16.4.0599
- Ashton, John; Ubido, Janet (1991). "The Healthy City and the Ecological Idea" (PDF). Journal of the Society for the Social History of Medicine. 4 (1): 173–181. doi:10.1093/shm/4.1.173. Retrieved 8 July 2013
- Tilley, David F. (2011). Aerobic Wastewater Treatment Processes: History and Development. IWA Publishing. ISBN 9781843395423. Retrieved 2013-02-07

Water Quality and Treatment: A Comprehensive Study

Water quality is the measurement of the condition of water. Water fluoridation, water purification, desalination, water treatment and reclaimed water are some of the topics discussed in the section. The aspects elucidated in this chapter are of vital importance, and provide a better understanding of water quality.

Water Quality

A rosette sampler is used to collect water samples in deep water, such as the Great Lakes or oceans, for water quality testing.

Water quality refers to the chemical, physical, biological, and radiological characteristics of water. It is a measure of the condition of water relative to the requirements of one or more biotic species and or to any human need or purpose. It is most frequently used by reference to a set of standards against which compliance can be assessed. The most common standards used to assess water quality relate to health of ecosystems, safety of human contact, and drinking water.

Standards

In the setting of standards, agencies make political and technical/scientific decisions about how the water will be used. In the case of natural water bodies, they also make some reasonable estimate of pristine conditions. Natural water bodies will vary in response to environmental conditions. Environmental scientists work to understand how these systems function, which in turn helps to identify the sources and fates of contaminants. Environmental lawyers and policymakers work to define legislation with the intention that water is maintained at an appropriate quality for its identified use.

The vast majority of surface water on the Earth is neither potable nor toxic. This remains true when seawater in the oceans (which is too salty to drink) is not counted. Another general perception of water quality is that of a simple property that tells whether water is polluted or not. In fact, water quality is a complex subject, in part because water is a complex medium intrinsically tied to the ecology of the Earth. Industrial and commercial activities (e.g. manufacturing, mining, construction, transport) are a major cause of water pollution as are runoff from agricultural areas, urban runoff and discharge of treated and untreated sewage.

Categories

The parameters for water quality are determined by the intended use. Work in the area of water quality tends to be focused on water that is treated for human consumption, industrial use, or in the environment.

Human Consumption

Contaminants that may be in untreated water include microorganisms such as viruses, protozoa and bacteria; inorganic contaminants such as salts and metals; organic chemical contaminants from industrial processes and petroleum use; pesticides and herbicides; and radioactive contaminants. Water quality depends on the local geology and ecosystem, as well as human uses such as sewage dispersion, industrial pollution, use of water bodies as a heat sink, and overuse (which may lower the level of the water).

The United States Environmental Protection Agency (EPA) limits the amounts of certain contaminants in tap water provided by US public water systems. The Safe Drinking Water Act authorizes EPA to issue two types of standards:

- *primary standards* regulate substances that potentially affect human health;
- *secondary standards* prescribe aesthetic qualities, those that affect taste, odor, or appearance.

The U.S. Food and Drug Administration (FDA) regulations establish limits for contaminants in bottled water that must provide the same protection for public health. Drinking water, including bottled water, may reasonably be expected to contain at least small amounts of some contaminants. The presence of these contaminants does not necessarily indicate that the water poses a health risk.

In urbanized areas around the world, water purification technology is used in municipal water systems to remove contaminants from the source water (surface water or groundwater) before it is distributed to homes, businesses, schools and other recipients. Water drawn directly from a stream, lake, or aquifer and that has no treatment will be of uncertain quality.

Industrial and Domestic Use

Dissolved minerals may affect suitability of water for a range of industrial and domestic purposes. The most familiar of these is probably the presence of ions of calcium (Ca^{2+}) and magnesium (Mg^{2+}) which interfere with the cleaning action of soap, and can form hard sulfate and soft carbonate deposits in water heaters or boilers. Hard water may be softened to remove these ions. The softening process often substitutes sodium cations. Hard water may be preferable to soft water for human consumption, since health problems have been associated with excess sodium and with calcium and magnesium deficiencies. Softening decreases nutrition and may increase cleaning effectiveness. Various industries' wastes and effluents can also pollute the water quality in receiving bodies of water.

Environmental Water Quality

Urban runoff discharging to coastal waters

Environmental water quality, also called ambient water quality, relates to water bodies such as lakes, rivers, and oceans. Water quality standards for surface waters vary significantly due to different environmental conditions, ecosystems, and intended human uses. Toxic substances and high populations of certain microorganisms can present a health hazard for non-drinking purposes such as irrigation, swimming, fishing, rafting, boating, and industrial uses. These conditions may also affect wildlife, which use the water for drinking or as a habitat. Modern water quality laws generally specify protection of fisheries and recreational use and require, as a minimum, retention of current quality standards.

Satirical cartoon by William Heath, showing a woman observing monsters in a drop of London water (at the time of the *Commission on the London Water Supply* report, 1828).

There is some desire among the public to return water bodies to pristine, or pre-industrial conditions. Most current environmental laws focus on the designation of particular uses of a water body. In some countries these designations allow for some water contamination as long as the particular type of contamination is not harmful to the designated uses. Given the landscape changes (e.g., land development, urbanization, clearcutting in forested areas) in the watersheds of many freshwater bodies, returning to pristine conditions would be a significant challenge. In these cases, environmental scientists focus on achieving goals for maintaining healthy ecosystems and may concentrate on the protection of populations of endangered species and protecting human health.

Sampling and Measurement

The complexity of water quality as a subject is reflected in the many types of measurements of water quality indicators. The most accurate measurements of water quality are made on-site, because water exists in equilibrium with its surroundings. Measurements commonly made on-site and in direct contact with the water source in question include temperature, pH, dissolved oxygen, conductivity, oxygen reduction potential (ORP), turbidity, and Secchi disk depth.

Sample Collection

More complex measurements are often made in a laboratory requiring a water sample to be collected, preserved, transported, and analyzed at another location. The process of water sampling introduces two significant problems:

- The first problem is the extent to which the sample may be representative of the water source of interest. Many water sources vary with time and with location. The measurement of interest may vary seasonally or from day to night or in response to some activity of man or natural populations of aquatic plants and animals. The measurement of interest may vary with distances from the water boundary with overlying atmosphere and underlying or confining soil.

The sampler must determine if a single time and location meets the needs of the investigation, or if the water use of interest can be satisfactorily assessed by averaged values with time and/or location, or if critical maxima and minima require individual measurements over a range of times, locations and/or events. The sample collection procedure must assure correct weighting of individual sampling times and locations where averaging is appropriate. Where critical maximum or minimum values exist, statistical methods must be applied to observed variation to determine an adequate number of samples to assess probability of exceeding those critical values.

- The second problem occurs as the sample is removed from the water source and begins to establish chemical equilibrium with its new surroundings – the sample container. Sample containers must be made of materials with minimal reactivity with substances to be measured; and pre-cleaning of sample containers is important. The water sample may dissolve part of the sample container and any residue on that container, or chemicals dissolved in the water sample may sorb onto the sample container and remain there when the water is poured out for analysis. Similar physical and chemical interactions may take place with any pumps, piping, or intermediate devices used to transfer the water sample into the sample container. Water collected from depths below the surface will normally be held at the reduced pressure of the atmosphere; so gas dissolved in the water may escape into unfilled space at the top of the container. Atmospheric gas present in that air space may also dissolve into the water sample. Other chemical reaction equilibria may change if the water sample changes temperature. Finely divided solid particles formerly suspended by water turbulence may settle to the bottom of the sample container, or a solid phase may form from biological growth or chemical precipitation. Microorganisms within the water sample may biochemically alter concentrations of oxygen, carbon dioxide, and organic compounds. Changing carbon dioxide concentrations may alter pH and change solubility of chemicals of interest. These problems are of special concern during measurement of chemicals assumed to be significant at very low concentrations.

Sample preservation may partially resolve the second problem. A common procedure is keeping samples cold to slow the rate of chemical reactions and phase change, and analyzing the sample as soon as possible; but this merely minimizes the changes rather than preventing them. A useful procedure for determining influence of sample containers during delay between sample collection and analysis involves preparation for two artificial samples in advance of the sampling event. One sample container is filled with water known from previous analysis to contain no detectable amount of the chemical of interest. This sample, called a "blank", is opened for exposure to the atmosphere when the sample of interest is collected, then resealed and transported to the laboratory with the sample for analysis to determine if sample holding procedures introduced any measurable amount of the chemical of interest. The second artificial sample is collected with the sample of interest, but then "spiked" with a measured additional amount of the chemical of interest at the time of collection. The blank and spiked samples are carried with the sample of interest and analyzed by the same methods at the same times to determine any changes indicating gains or losses during the elapsed time between collection and analysis.

An automated sampling station installed along the East Branch Milwaukee River, New Fane, Wisconsin. The cover of the 24-bottle autosampler (center) is partially raised, showing the sample bottles inside. The autosampler was programmed to collect samples at time intervals, or proportionate to flow over a specified period. The data logger (white cabinet) recorded temperature, specific conductance,and dissolved oxygen levels.

Filtering a manually collected water sample (grab sample) for analysis.

Testing in Response to Emergencies

Inevitably after events such as earthquakes and tsunamis, there is an immediate response by the aid agencies as relief operations get underway to try and restore basic infrastructure and provide the basic fundamental items that are necessary for survival and subsequent recovery. Access to clean drinking water and adequate sanitation is a priority at times like this. The threat of disease increases hugely due to the large numbers of people living close together, often in squalid conditions, and without proper sanitation.

After a natural disaster, as far as water quality testing is concerned there are widespread views on the best course of action to take and a variety of methods can be employed. The key basic water quality parameters that need to be addressed in an emergency are bacteriological indicators of fecal contamination, free chlorine residual, pH, turbidity and possibly conductivity/total dissolved solids. There are a number of portable water test kits on the market widely used by aid and relief agencies for carrying out such testing.

After major natural disasters, a considerable length of time might pass before water quality returns to pre-disaster levels. For example, following the 2004 Indian Ocean tsunami the Colombo-based International Water Management Institute (IWMI) monitored the effects of saltwater and concluded that the wells recovered to pre-tsunami drinking water quality one and a half years after the event. IWMI developed protocols for cleaning wells contaminated by saltwater; these were subsequently officially endorsed by the World Health Organization as part of its series of Emergency Guidelines.

Chemical Analysis

A gas chromatograph-
mass spectrometer measures pesticides and other organic pollutants.

The simplest methods of chemical analysis are those measuring chemical elements without respect to their form. Elemental analysis for oxygen, as an example, would indicate a concentration of 890,000 milligrams per litre (mg/L) of water sample because water is made of oxygen. The method selected to measure dissolved oxygen should differentiate between diatomic oxygen and oxygen combined with other elements. The comparative simplicity of elemental analysis has produced a large amount of sample data and water quality criteria for elements sometimes identified as heavy metals. Water analysis for heavy metals must consider soil particles suspended in the water sample. These suspended soil particles may contain measurable amounts of metal. Although the particles are not dissolved in the water, they may be consumed by people drinking the water. Adding acid to a water sample to prevent loss of dissolved metals onto the sample container may dissolve more metals from suspended soil particles. Filtration of soil particles from the water sample before acid addition, however, may cause loss of dissolved metals onto the filter. The complexities of differentiating similar organic molecules are even more challenging.

Making these complex measurements can be expensive. Because direct measurements of water quality can be expensive, ongoing monitoring programs are typically conducted by government agencies. However, there are local volunteer programs and resources available for some general assessment. Tools available to the general public include on-site test kits, commonly used for home fish tanks, and biological assessment procedures.

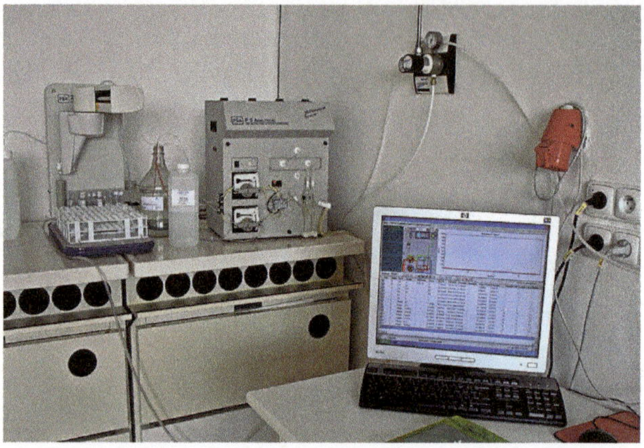

Atomic fluorescence spectroscopy is used to measure mercury and other heavy metals.

Real-time Monitoring

Although water quality is usually sampled and analyzed at laboratories, nowadays, citizens demand real-time information about the water they are drinking. During the last years, several companies are deploying worldwide real-time remote monitoring systems for measuring water pH, turbidity or dissolved oxygen levels.

Drinking Water Indicators

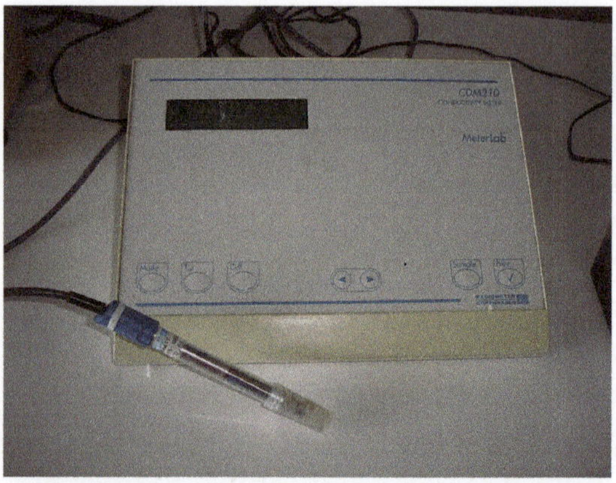

An electrical conductivity meter is used to measure total dissolved solids.

The following is a list of indicators often measured by situational category:

- Alkalinity
- Color of water
- pH

- Taste and odor (geosmin, 2-Methylisoborneol (MIB), etc.)
- Dissolved metals and salts (sodium, chloride, potassium, calcium, manganese, magnesium)
- Microorganisms such as fecal coliform bacteria (*Escherichia coli*), Cryptosporidium, and Giardia lamblia
- Dissolved metals and metalloids (lead, mercury, arsenic, etc.)
- Dissolved organics: colored dissolved organic matter (CDOM), dissolved organic carbon (DOC)
- Radon
- Heavy metals
- Pharmaceuticals
- Hormone analogs

Environmental Indicators

Physical Indicators

- Water temperature
- Specific conductance or electrical conductance (EC) or conductivity
- Total suspended solids (TSS)
- Transparency or turbidity
- Total dissolved solids (TDS)
- Odour of water
- Color of water
- Taste of water

Chemical Indicators

- pH
- Biochemical oxygen demand (BOD)
- Chemical oxygen demand (COD)
- Dissolved oxygen (DO)
- Total hardness (TH)
- Heavy metals
- Nitrate
- Orthophosphates
- Pesticides
- Surfactants

Biological Indicators

Biological monitoring metrics have been developed in many places, and one widely used measure is the presence and abundance of members of the insect orders Ephemeroptera, Plecoptera and Trichoptera (common names are, respectively, mayfly, stonefly and caddisfly). EPT indexes will naturally vary from region to region, but generally, within a region, the greater the number of taxa from these orders, the better the water quality. Organisations in the United States, such as EPA offer guidance on developing a monitoring program and identifying members of these and other aquatic insect orders.

- Ephemeroptera
- Plecoptera
- Mollusca
- Trichoptera
- *Escherichia coli* (E. coli)
- Coliform bacteria

Individuals interested in monitoring water quality who cannot afford or manage lab scale analysis can also use biological indicators to get a general reading of water quality. One example is the IOWATER volunteer water monitoring program, which includes a benthic macroinvertebrate indicator key.

Bivalve molluscs are largely used as bioindicators to monitor the health of aquatic environments in both fresh water and the marine environments. Their population status or structure, physiology, behaviour or the level of contamination with elements or compounds can indicate the state of contamination status of the ecosystem. They are particularly useful since they are sessile so that they are representative of the environment where they are sampled or placed. A typical project is the Mussel Watch Programme, but today they are used worldwide.

The Southern African Scoring System (SASS) method is a biological water quality monitoring system based on the presence of benthic macroinvertebrates. The SASS aquatic biomonitoring tool has been refined over the past 30 years and is now on the fifth version (SASS5) which has been specifically modified in accordance with international standards, namely the ISO/IEC 17025 protocol. The SASS5 method is used by the South African Department of Water Affairs as a standard method for River Health Assessment, which feeds the national River Health Programme and the national Rivers Database.

Standards and Reports

International

- The World Health Organisation (WHO) has published guidelines for drinking-water quality (GDWQ) in 2011.
- The International Organization for Standardization (ISO) published regulation of water quality in the section of ICS 13.060, ranging from water sampling, drinking water, industrial class water, sewage, and examination of water for chemical, physical or biological properties. ICS 91.140.60 covers the standards of water supply systems.

National Specifications for Ambient Water and Drinking water

European Union

The water policy of the European Union is primarily codified in three directives:

- Directive on Urban Waste Water Treatment (91/271/EEC) of 21 May 1991 concerning discharges of municipal and some industrial wastewaters;
- The Drinking Water Directive (98/83/EC) of 3 November 1998 concerning potable water quality;
- Water Framework Directive (2000/60/EC) of 23 October 2000 concerning water resources management.

India

- In an Council of Medical Research (ICMR) Standards for Drinking Water.

South Africa

Water quality guidelines for South Africa are grouped according to potential user types (e.g. domestic, industrial) in the 1996 Water Quality Guidelines. Drinking water quality is subject to the South African National Standard (SANS) 241 Drinking Water Specification.

United Kingdom

In England and Wales acceptable levels for drinking water supply are listed in the "Water Supply (Water Quality) Regulations 2000."

United States

In the United States, Water Quality Standards are defined by state agencies for various water bodies, guided by the desired uses for the water body (e.g., fish habitat, drinking water supply, recreational use). The Clean Water Act (CWA) requires each governing jurisdiction (states, territories, and covered tribal entities) to submit a set of biennial reports on the quality of water in their area. These reports are known as the 303(d) and 305(b) reports, named for their respective CWA provisions, and are submitted to, and approved by, EPA. These reports are completed by the governing jurisdiction, typically a state environmental agency. EPA recommends that each state submit a single "Integrated Report" comprising its list of impaired waters and the status of all water bodies in the state. The *National Water Quality Inventory Report to Congress* is a general report on water quality, providing overall information about the number of miles of streams and rivers and their aggregate condition. The CWA requires states to adopt standards for each of the possible designated uses that they assign to their waters. Should evidence suggest or document that a stream, river or lake has failed to meet the water quality criteria for one or more of its designated uses, it is placed on a list of impaired waters. Once a state has placed a water body on this list, it must develop a management plan establishing Total Maximum Daily Loads (TMDLs) for the pollutant(s) impairing the use of the water. These TMDLs establish the reductions needed to fully support the designated uses.

Drinking water standards, which are applicable to public water systems, are issued by EPA under the Safe Drinking Water Act.

Water Fluoridation

Fluoridation does not affect the appearance, taste or smell of drinking water.

Water fluoridation is the controlled addition of fluoride to a public water supply to reduce tooth decay. Fluoridated water has fluoride at a level that is effective for preventing cavities; this can occur naturally or by adding fluoride. Fluoridated water operates on tooth surfaces: in the mouth, it creates low levels of fluoride in saliva, which reduces the rate at which tooth enamel demineralizes and increases the rate at which it remineralizes in the early stages of cavities. Typically a fluoridated compound is added to drinking water, a process that in the U.S. costs an average of about $1.04 per person-year. Defluoridation is needed when the naturally occurring fluoride level exceeds recommended limits. In 2011 the World Health Organization suggested a level of fluoride from 0.5 to 1.5 mg/L (milligrams per litre), depending on climate, local environment, and other sources of fluoride. Bottled water typically has unknown fluoride levels.

Dental caries remains a major public health concern in most industrialized countries, affecting 60–90% of schoolchildren and the vast majority of adults. Water fluoridation reduces cavities in children, while efficacy in adults is less clear. A Cochrane review estimates a reduction in cavities when water fluoridation was used by children who had no access to other sources of fluoride to be 35% in baby teeth and 26% in permanent teeth. The evidence quality was poor. Most European countries have experienced substantial declines in tooth decay without its use. Recent studies suggest that water fluoridation, particularly in industrialized nations, may be unnecessary because topical fluorides (such as in toothpaste) are widely used, and caries rates have become low.

Although fluoridation can cause dental fluorosis, which can alter the appearance of developing teeth or enamel fluorosis, the differences are mild and usually not considered to be of aesthetic or public health concern. There is no clear evidence of other adverse effects from water fluoridation. Fluoride's effects depend on the total daily intake of fluoride from all sources. Drinking water is typically the largest source; other methods of fluoride therapy include fluoridation of toothpaste, salt, and milk. The views on the most efficient method for community prevention of tooth decay are mixed. The Australian government states that water fluoridation is the most effective way to achieve fluoride exposure that is community-wide. The World Health Organization reports that water fluoridation, when feasible and culturally acceptable, has substantial advantages, especially for subgroups at high risk, while the European Commission finds no benefit to water fluoridation compared with topical use.

Public water fluoridation was first practiced in the U.S. As of 2012, 25 countries have artificial water fluoridation to varying degrees, 11 of them have more than 50% of their population drinking fluoridated water. A further 28 countries have water that is naturally fluoridated, though in many of them the fluoride is above the recommended safe level. As of 2012, about 435 million people worldwide received water fluoridated at the recommended level (i.e., about 5.4% of the global population). About 214 million of them living in the United States. Major health organizations such as the World Health Organization and FDI World Dental Federation supported water fluoridation as safe and effective. The Centers for Disease Control and Prevention lists water fluoridation as one of the ten great public health achievements of the 20th century in the U.S. Despite this, the practice is controversial as a public health measure; some countries and communities have discontinued it, while others have expanded it. Opponents of the practice argue that neither the benefits nor the risks have been studied adequately, and debate the conflict between what might be considered mass medication and individual liberties.

Goal

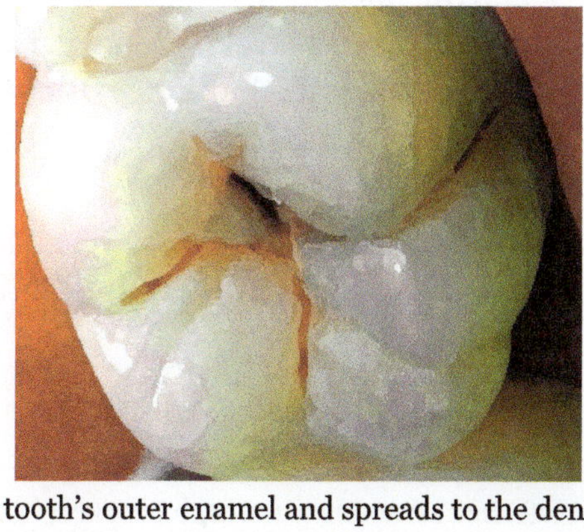

A cavity starts in a tooth's outer enamel and spreads to the dentin and pulp inside.

The goal of water fluoridation is to prevent tooth decay by adjusting the concentration of fluoride in public water supplies. Tooth decay (dental caries) is one of the most prevalent chronic diseases worldwide. Although it is rarely life-threatening, tooth decay can cause pain and impair eating, speaking, facial appearance, and acceptance into society, and it greatly affects the quality of life of children, particularly those of low socioeconomic status. In most industrialized countries, tooth decay affects 60–90% of schoolchildren and the vast majority of adults; although the problem appears to be less in Africa's developing countries, it is expected to increase in several countries there because of changing diet and inadequate fluoride exposure. In the U.S., minorities and the poor both have higher rates of decayed and missing teeth, and their children have less dental care. Once a cavity occurs, the tooth's fate is that of repeated restorations, with estimates for the median life of an amalgam tooth filling ranging from 9 to 14 years. Oral disease is the fourth most expensive disease to treat. The motivation for fluoridation of salt or water is similar to that of iodized salt for the prevention of mental retardation and goiter.

The goal of water fluoridation is to prevent a chronic disease whose burdens particularly fall on children and the poor. Another of the goals was to bridge inequalities in dental health and dental care. Some studies suggest that fluoridation reduces oral health inequalities between the rich and poor, but the evidence is limited. There is anecdotal but not scientific evidence that fluoride allows more time for dental treatment by slowing the progression of tooth decay, and that it simplifies treatment by causing most cavities to occur in pits and fissures of teeth. Other reviews have found not enough evidence to determine if water fluoridation reduces oral-health social disparities.

Its use presents a conflict between the common good and individual rights. It is controversial, and opposition to it has been based on ethical, legal, safety, and efficacy grounds. Health and dental organizations worldwide have endorsed its safety and effectiveness. Its use began in 1945, following studies of children in a region where higher levels of fluoride occur naturally in the water. Further research showed that moderate fluoridation prevents tooth decay.

Implementation

Fluoride monitor (at left) in a community water tower pumphouse, Minnesota, 1987

Fluoridation does not affect the appearance, taste, or smell of drinking water. It is normally accomplished by adding one of three compounds to the water: sodium fluoride, fluorosilicic acid, or sodium fluorosilicate.

- Sodium fluoride (NaF) was the first compound used and is the reference standard. It is a white, odorless powder or crystal; the crystalline form is preferred if manual handling is used, as it minimizes dust. It is more expensive than the other compounds, but is easily handled and is usually used by smaller utility companies. It is toxic in gram quantities by ingestion or inhalation.
- Fluorosilicic acid (H_2SiF_6) is the most commonly used additive for water fluoridation in the United States. It is an inexpensive liquid by-product of phosphate fertilizer manufacture. It comes in varying strengths, typically 23–25%; because it contains so much water, shipping can be expensive. It is also known as hexafluorosilicic, hexafluosilicic, hydrofluosilicic, and silicofluoric acid.
- Sodium fluorosilicate (Na_2SiF_6) is the sodium salt of fluorosilicic acid. It is a powder or very fine crystal that is easier to ship than fluorosilicic acid. It is also known as sodium silicofluoride.

These compounds were chosen for their solubility, safety, availability, and low cost. A 1992 census found that, for U.S. public water supply systems reporting the type of compound used, 63% of the population received water fluoridated with fluorosilicic acid, 28% with sodium fluorosilicate, and 9% with sodium fluoride.

Recommendations

The Centers for Disease Control and Prevention developed recommendations for water fluoridation that specify requirements for personnel, reporting, training, inspection, monitoring, surveillance, and actions in case of overfeed, along with technical requirements for each major compound used.

Although fluoride was once considered an essential nutrient, the U.S. National Research Council has since removed this designation due to the lack of studies showing it is essential for human growth, though still considering fluoride a "beneficial element" due to its positive impact on oral health. The European Food Safety Authority's Panel on Dietetic Products, Nutrition and Allergies (NDA) considers fluoride not to be an essential nutrient, yet, due to the beneficial effects of dietary fluoride on prevention of dental caries they have defined an Adequate Intake (AI) value for it. The AI of fluoride from all sources (including non-dietary sources) is 0.05 mg/kg body weight per day for both children and adults, including pregnant and lactating women.

In 2011, the U.S. Department of Health and Human Services (HHS) and the U.S. Environmental Protection Agency (EPA) lowered the recommended level of fluoride to 0.7 mg/L. In 2015, the U.S. Food and Drug Administration (FDA), based on the recommendation of the U.S. Public Health Service (PHS) for fluoridation of community water systems, recommended that bottled water manufacturers limit fluoride in bottled water to no more than 0.7 milligrams per liter (mg/L)(milligrams per liter, equivalent to parts per million).

Previous recommendations were based on evaluations from 1962, when the U.S. specified the optimal level of fluoride to range from 0.7 to 1.2 mg/L (milligrams per liter, equivalent to parts per million), depending on the average maximum daily air temperature; the optimal level is lower in warmer climates, where people drink more water, and is higher in cooler climates.

These standards are not appropriate for all parts of the world, where fluoride levels might be excessive and fluoride should be removed from water, and is based on assumptions that have become obsolete with the rise of air conditioning and increased use of soft drinks, processed food, fluoridated toothpaste, and other sources of fluorides. In 2011 the World Health Organization stated that 1.5 mg/L should be an absolute upper bound and that 0.5 mg/L may be an appropriate lower limit. A 2007 Australian systematic review recommended a range from 0.6 to 1.1 mg/L.

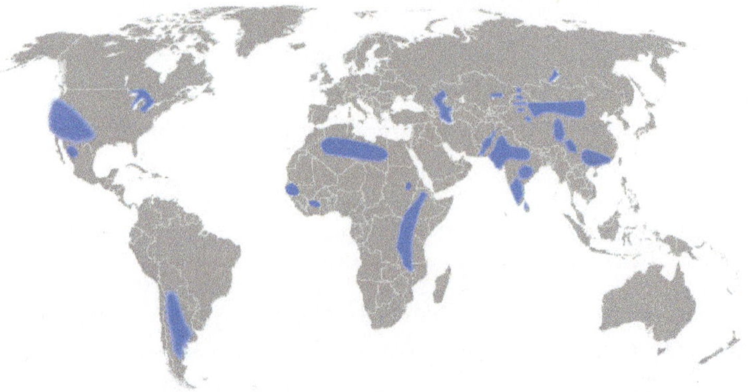

Geographical areas associated with groundwater having over 1.5 mg/L of naturally occurring fluoride, which is above recommended levels.

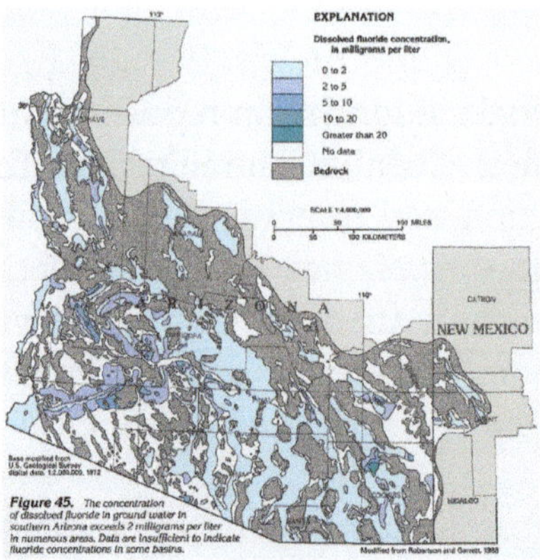

Detail of southern Arizona. Areas in darker blues have groundwater with over 2 mg/L of naturally occurring fluoride.

Occurrences

Fluoride naturally occurring in water can be above, at, or below recommended levels. Rivers and lakes generally contain fluoride levels less than 0.5 mg/L, but groundwater, particularly in volcanic or mountainous areas, can contain as much as 50 mg/L. Higher concentrations of fluorine are found in alkaline volcanic, hydrothermal, sedimentary, and other rocks derived from highly evolved magmas and hydrothermal solutions, and this fluorine dissolves into nearby water as fluoride. In most drinking waters, over 95% of total fluoride is the F^- ion, with the magnesium–fluoride complex (MgF^+) being the next most common. Because fluoride levels in water are usually controlled by the solubility of fluorite (CaF_2), high natural fluoride levels are associated with calcium-deficient, alkaline, and soft waters. Defluoridation is needed when the naturally occurring fluoride level exceeds recommended limits. It can be accomplished by percolating water through granular beds of activated alumina, bone meal, bone char, or tricalcium phosphate; by coagulation with alum; or by precipitation with lime.

Pitcher or faucet-mounted water filters do not alter fluoride content; the more-expensive reverse osmosis filters remove 65–95% of fluoride, and distillation removes all fluoride. Some bottled waters contain undeclared fluoride, which can be present naturally in source waters, or if water is sourced from a public supply which has been fluoridated. The FDA states that bottled water products labeled as de-ionized, purified, demineralized, or distilled have been treated in such a way that they contain no or only trace amounts of fluoride, unless they specifically list fluoride as an added ingredient.

Evidence

Existing evidence suggests that water fluoridation reduces tooth decay. Consistent evidence also suggests that it causes dental fluorosis, most of which is mild and not usually of aesthetic concern. No clear evidence of other adverse effects exists, though almost all research thereof has been of poor quality.

Effectiveness

Reviews have shown that water fluoridation reduces cavities in children. A conclusion for the efficacy in adults is less clear with some reviews finding benefit and others not. Studies in the U.S. in the 1950s and 1960s showed that water fluoridation reduced childhood cavities by fifty to sixty percent, while studies in 1989 and 1990 showed lower reductions (40% and 18% respectively), likely due to increasing use of fluoride from other sources, notably toothpaste, and also the 'halo effect' of food and drink that is made in fluoridated areas and consumed in unfluoridated ones.

A 2000 UK systematic review (York) found that water fluoridation was associated with a decreased proportion of children with cavities of 15% and with a decrease in decayed, missing, and filled primary teeth (average decreases was 2.25 teeth). The review found that the evidence was of moderate quality: few studies attempted to reduce observer bias, control for confounding factors, report variance measures, or use appropriate analysis. Although no major differences between natural and artificial fluoridation were apparent, the evidence was inadequate for a conclusion about any differences. A 2007 Australian systematic review used the same inclusion criteria as York's, plus one additional study. This did not affect the York conclusions. A 2011 European Commission systematic review based its efficacy on York's review conclusion. A 2015 Cochrane systematic review estimated a reduction in cavities when water fluoridation was used by children who had no access to other sources of fluoride to be 35% in baby teeth and 26% in permanent teeth. The evidence was of poor quality.

Fluoride may also prevent cavities in adults of all ages. A 2007 meta-analysis by CDC researchers found that water fluoridation prevented an estimated 27% of cavities in adults, about the same fraction as prevented by exposure to any delivery method of fluoride (29% average). A 2011 European Commission review found that the benefits of water fluoridation for adult in terms of reductions in decay are limited. A 2015 Cochrane review found no conclusive research regarding the effectiveness of water fluoridation in adults. A 2016 review found variable quality evidence that, overall, stopping of community water fluoridation programs was typically followed by an increase in cavities.

Most countries in Europe have experienced substantial declines in cavities without the use of water fluoridation. For example, in Finland and Germany, tooth decay rates remained stable or continued to decline after water fluoridation stopped. Fluoridation may be useful in the U.S. because unlike most European countries, the U.S. does not have school-based dental care, many children do not visit a dentist regularly, and for many U.S. children water fluoridation is the prime source of exposure to fluoride. The effectiveness of water fluoridation can vary according to circumstances such as whether preventive dental care is free to all children.

Fluorosis

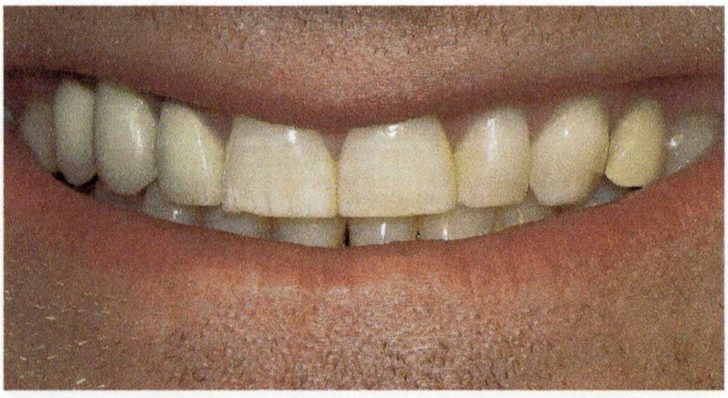

A mild case of dental fluorosis, visible as white streaks on the subject's upper right central incisor.

Fluoride's adverse effects depend on total fluoride dosage from all sources. At the commonly recommended dosage, the only clear adverse effect is dental fluorosis, which can alter the appearance of children's teeth during tooth development; this is mostly mild and is unlikely to represent any real effect on aesthetic appearance or on public health. In April 2015, recommended fluoride levels in the United States were changed to 0.7 ppm from 0.7–1.2 ppm to reduce the risk of dental fluorosis. The 2015 Cochrane review estimated that for a fluoride level of 0.7 ppm the percentage of participants with fluorosis of aesthetic concern was approximately 12%. This increases to 40% when considering fluorosis of any level not of aesthetic concern. In the US mild or very mild dental fluorosis has been reported in 20% of the population, moderate fluorosis in 2% and severe fluorosis in less than 1%.

The critical period of exposure is between ages one and four years, with the risk ending around age eight. Fluorosis can be prevented by monitoring all sources of fluoride, with fluoridated water directly or indirectly responsible for an estimated 40% of risk and other sources, notably toothpaste, responsible for the remaining 60%. Compared to water naturally fluoridated at 0.4 mg/L, fluoridation to 1 mg/L is estimated to cause additional fluorosis in one of every 6 people (95% CI 4–21 people), and to cause additional fluorosis of aesthetic concern in one of every 22 people (95% CI 13.6–∞ people). Here, *aesthetic concern* is a term used in a standardized scale based on what adolescents would find unacceptable, as measured by a 1996 study of British 14-year-olds. In many industrialized countries the prevalence of fluorosis is increasing even in unfluoridated communities, mostly because of fluoride from swallowed toothpaste. A 2009 systematic review indicated that fluorosis is associated with consumption of infant formula or of water added to reconstitute the formula, that the evidence was distorted by publication bias, and that the evidence that the formula's fluoride caused the fluorosis was weak. In the U.S. the decline in tooth decay was accompanied by increased fluorosis in both fluoridated and unfluoridated communities; accordingly, fluoride has been reduced in various ways worldwide in infant formulas, children's toothpaste, water, and fluoride-supplement schedules.

Safety

Fluoridation has little effect on risk of bone fracture (broken bones); it may result in slightly lower fracture risk than either excessively high levels of fluoridation or no fluoridation. There is no clear association between fluoridation and cancer or deaths due to cancer, both for cancer in general and also specifically for bone cancer and osteosarcoma. Other adverse effects lack sufficient evidence to reach a confident conclusion.

Fluoride can occur naturally in water in concentrations well above recommended levels, which can have several long-term adverse effects, including severe dental fluorosis, skeletal fluorosis, and weakened bones; water utilities in the developed world reduce fluoride levels to regulated maximum levels in regions where natural levels are high, and the WHO and other groups work with countries and regions in the developing world with naturally excessive fluoride levels to achieve safe levels. The World Health Organization recommends a guideline maximum fluoride value of 1.5 mg/L as a level at which fluorosis should be minimal.

In rare cases improper implementation of water fluoridation can result in overfluoridation that causes outbreaks of acute fluoride poisoning, with symptoms that include nausea, vomiting, and diarrhea. Three such outbreaks were reported in the U.S. between 1991 and 1998, caused by fluoride concentrations as high as 220 mg/L; in the 1992 Alaska outbreak, 262 people became ill and one person died. In 2010, approximately 60 gallons of fluoride were released into the water supply in Asheboro, North Carolina in 90 minutes—an amount that was intended to be released in a 24-hour period.

Like other common water additives such as chlorine, hydrofluosilicic acid and sodium silicofluoride decrease pH and cause a small increase of corrosivity, but this problem is easily addressed by increasing the pH. Although it has been hypothesized that hydrofluosilicic acid and sodium silicofluoride might increase human lead uptake from water, a 2006 statistical analysis did not support concerns that these chemicals cause higher blood lead concentrations in children. Trace levels of arsenic and lead may be present in fluoride compounds added to water, but no credible evidence exists that their presence is of concern: concentrations are below measurement limits.

The effect of water fluoridation on the natural environment has been investigated, and no adverse effects have been established. Issues studied have included fluoride concentrations in groundwater and downstream rivers; lawns, gardens, and plants; consumption of plants grown in fluoridated water; air emissions; and equipment noise.

Mechanism

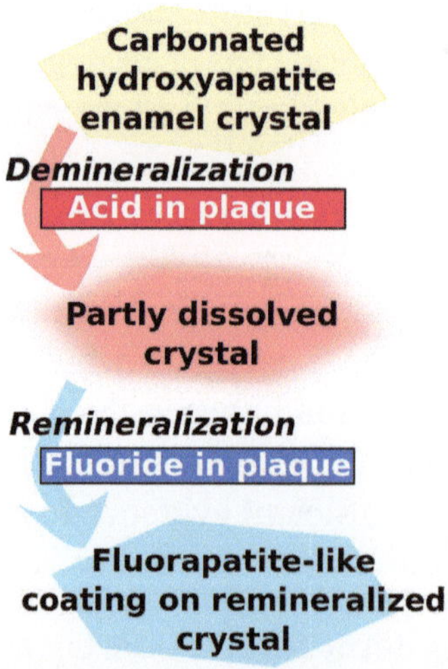

Demineralization and remineralization of dental enamel in the presence of acid and fluoride in saliva and plaque fluid.

Fluoride exerts its major effect by interfering with the demineralization mechanism of tooth decay. Tooth decay is an infectious disease, the key feature of which is an increase within dental plaque of bacteria such as *Streptococcus mutans* and *Lactobacillus*. These produce organic acids when carbohydrates, especially sugar, are eaten. When enough acid is produced to lower the pH below 5.5, the acid dissolves carbonated hydroxyapatite, the main component of tooth enamel, in a process known as *demineralization*. After the sugar is gone, some of the mineral loss can be recovered—or *remineralized*—from ions dissolved in the saliva. Cavities result when the rate of demineralization exceeds the rate of remineralization, typically in a process that requires many months or years.

All fluoridation methods, including water fluoridation, create low levels of fluoride ions in saliva and plaque fluid, thus exerting a topical or surface effect. A person living in an area with fluoridated water may experience rises of fluoride concentration in saliva to about 0.04 mg/L several times during a day. Technically, this fluoride does not prevent cavities but rather controls the rate at which they develop. When fluoride ions are present in plaque fluid along with dissolved hydroxyapatite, and the pH is higher than 4.5, a fluorapatite-like remineralized veneer is formed over the remaining surface of the enamel; this veneer is much more acid-resistant than the original hydroxyapatite, and is formed more quickly than ordinary remineralized enamel would be. The cavity-prevention effect of fluoride is mostly due to these surface effects, which occur during and after tooth eruption. Although some systemic (whole-body) fluoride returns to the saliva via blood plasma, and to unerupted teeth via plasma or crypt fluid, there is little data to determine what percentages of fluoride's anticavity effect comes from these systemic mechanisms. Also, although fluoride affects the physiology of dental bacteria, its effect on bacterial growth does not seem to be relevant to cavity prevention.

Fluoride's effects depend on the total daily intake of fluoride from all sources. About 70–90% of ingested fluoride is absorbed into the blood, where it distributes throughout the body. In infants 80–90% of absorbed fluoride is retained, with the rest excreted, mostly via urine; in adults about 60% is retained. About 99% of retained fluoride is stored in bone, teeth, and other calcium-rich areas, where excess quantities can cause fluorosis. Drinking water is typically the largest source of fluoride. In many industrialized countries swallowed toothpaste is the main source of fluoride exposure in unfluoridated communities. Other sources include dental products other than toothpaste; air pollution from fluoride-containing coal or from phosphate fertilizers; trona, used to tenderize meat in Tanzania; and tea leaves, particularly the tea bricks favored in parts of China. High fluoride levels have been found in other foods, including barley, cassava, corn, rice, taro, yams, and fish protein concentrate. The U.S. Institute of Medicine has established Dietary Reference Intakes for fluoride: Adequate Intake values range from 0.01 mg/day for infants aged 6 months or less, to 4 mg/day for men aged 19 years and up; and the Tolerable Upper Intake Level is 0.10 mg/kg/day for infants and children through age 8 years, and 10 mg/day thereafter. A rough estimate is that an adult in a temperate climate consumes 0.6 mg/day of fluoride without fluoridation, and 2 mg/day with fluoridation. However, these values differ greatly among the world's regions: for example, in Sichuan, China the average daily fluoride intake is only 0.1 mg/day in drinking water but 8.9 mg/day in food and 0.7 mg/day directly from the air due to the use of high-fluoride soft coal for cooking and drying foodstuffs indoors.

Alternatives

The views on the most effective method for community prevention of tooth decay are mixed. The Australian government review states that water fluoridation is the most effective means of achieving fluoride exposure that is community-wide. The European Commission review states "No obvious advantage appears in favour of water fluoridation compared with topical prevention". Other fluoride therapies are also effective in preventing tooth decay; they include fluoride toothpaste, mouthwash, gel, and varnish, and fluoridation of salt and milk. Dental sealants are effective as well, with estimates of prevented cavities ranging from 33% to 86%, depending on age of sealant and type of study.

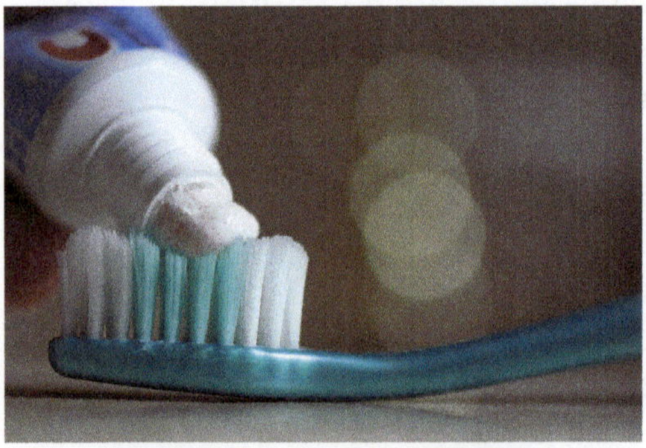

Fluoride toothpaste is effective against cavities. It is widely used, but less so among the poor.

Fluoridated iodized salt sold in Germany

Fluoride toothpaste is the most widely used and rigorously evaluated fluoride treatment. Its introduction in the early 1970s is considered the main reason for the decline in tooth decay in industrialized countries, and toothpaste appears to be the single common factor in countries where tooth decay has declined. Toothpaste is the only realistic fluoride strategy in many low-income countries, where lack of infrastructure renders water or salt fluoridation infeasible. It relies on individual and family behavior, and its use is less likely among lower economic classes; in low-income countries it is unaffordable for the poor. Fluoride toothpaste prevents about 25% of cavities in young permanent teeth, and its effectiveness is improved if higher concentrations of fluoride are used, or if the toothbrushing is supervised. Fluoride mouthwash and gel are about as effective as fluoride toothpaste; fluoride varnish prevents about 45% of cavities. By comparison, brushing with a nonfluoride toothpaste has little effect on cavities.

The effectiveness of salt fluoridation is about the same as that of water fluoridation, if most salt for human consumption is fluoridated. Fluoridated salt reaches the consumer in salt at home, in meals at school and at large kitchens, and in bread. For example, Jamaica has just one salt producer, but a complex public water supply; it started fluoridating all salt in 1987, achieving a decline in cavities. Universal salt fluoridation is also practiced in Colombia and the Swiss Canton of Vaud; in Germany fluoridated salt is widely used in households but unfluoridated salt is also available, giving consumers a choice. Concentrations of fluoride in salt range from 90 to 350 mg/kg, with studies suggesting an optimal concentration of around 250 mg/kg.

Milk fluoridation is practiced by the Borrow Foundation in some parts of Bulgaria, Chile, Peru, Russia, Macedonia, Thailand and the UK. Depending on location, the fluoride is added to milk, to powdered milk, or to yogurt. For example, milk powder fluoridation is used in rural Chilean areas where water fluoridation is not technically feasible. These programs are aimed at children, and have neither targeted nor been evaluated for adults. A systematic review found low-quality evidence to support the practice, but also concluded that further studies were needed.

Other public-health strategies to control tooth decay, such as education to change behavior and diet, have lacked impressive results. Although fluoride is the only well-documented agent which controls the rate at which cavities develop, it has been suggested that adding calcium to the water would reduce cavities further. Other agents to prevent tooth decay include antibacterials such as chlorhexidine and sugar substitutes such as xylitol. Xylitol-sweetened chewing gum has been recommended as a supplement to fluoride and other conventional treatments if the gum is not too costly. Two proposed approaches, bacteria replacement therapy (probiotics) and caries vaccine, would share water fluoridation's advantage of requiring only minimal patient compliance, but have not been proven safe and effective. Other experimental approaches include fluoridated sugar, polyphenols, and casein phosphopeptide–amorphous calcium phosphate nanocomplexes.

A 2007 Australian review concluded that water fluoridation is the most effective and socially the most equitable way to expose entire communities to fluoride's cavity-prevention effects. A 2002 U.S. review estimated that sealants decreased cavities by about 60% overall, compared to about 18–50% for fluoride. A 2007 Italian review suggested that water fluoridation may not be needed, particularly in the industrialized countries where cavities have become rare, and concluded that toothpaste and other topical fluoride are the best way to prevent cavities worldwide. A 2004 World Health Organization review stated that water fluoridation, when it is culturally acceptable and technically feasible, has substantial advantages in preventing tooth decay, especially for subgroups at high risk.

Usage

As of November 2012, a total of about 378 million people worldwide received artificially fluoridated water. The majority of those were in the United States. About 40 million worldwide received water that was naturally fluoridated to recommended levels.

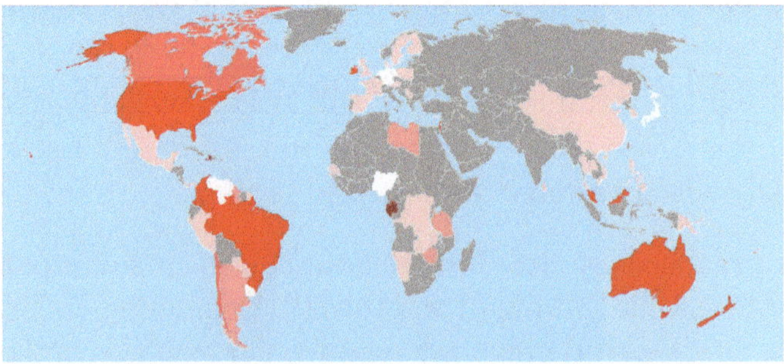

Percentage of population receiving fluoridated water, including both artificial and natural fluoridation.

80–100% 60–80% 40–60% 20–40% 1–20% < 1% unknown

Much of the early work on establishing the connection between fluoride and dental health was performed by scientists in the U.S. during the early 20th century, and the U.S. was the first country to implement public water fluoridation on a wide scale. It has been introduced to varying degrees in many countries and territories outside the U.S., including Argentina, Australia, Brazil, Canada, Chile, Colombia, Hong Kong, Ireland, Israel, Korea, Malaysia, New Zealand, the Philippines, Serbia, Singapore, Spain, the UK, and Vietnam. In 2004, an estimated 13.7 million people in western Europe and 194 million in the U.S. received artificially fluoridated water. In 2010, about 66% of the U.S. population was receiving fluoridated water.

Naturally fluoridated water is used by approximately 4% of the world's population, in countries including Argentina, France, Gabon, Libya, Mexico, Senegal, Sri Lanka, Tanzania, the U.S., and Zimbabwe. In some locations, notably parts of Africa, China, and India, natural fluoridation exceeds recommended levels.

Communities have discontinued water fluoridation in some countries, including Finland, Germany, Japan, the Netherlands, Sweden, and Switzerland. On August 26, 2014, Israel stopped mandating fluoridation, stating "Only some 1% of the water is used for drinking, while 99% of the water is intended for other uses (industry, agriculture, flushing toilets etc.). There is also scientific evidence that fluoride in large amounts can lead to damage to health. When fluoride is supplied via drinking water, there is no control regarding the amount of fluoride actually consumed, which could lead to excessive consumption. Supply of fluoridated water forces those who do not so wish to also consume water with added fluoride. This approach is therefore not accepted in most countries in the world." This change was often motivated by political opposition to water fluoridation, but sometimes the need for water fluoridation was met by alternative strategies. The use of fluoride in its various forms is the foundation of tooth decay prevention throughout Europe; several countries have introduced fluoridated salt, with varying success: in Switzerland and Germany, fluoridated salt represents 65% to 70% of the domestic market, while in France the market share reached 60% in 1993 but dwindled to 14% in 2009; Spain, in 1986 the second West European country to introduce fluoridation of table salt, reported a market share in 2006 of only 10%. In three other West European countries, Greece, Austria and the Netherlands, the legal framework for production and marketing of fluoridated edible salt exists. At least six Central European countries (Hungary, the Czech and Slovak Republics, Croatia, Slovenia, Romania) have shown some interest in salt fluoridation; however, significant usage of approximately 35% was only achieved in the Czech Republic. The Slovak Republic had the equipment to treat salt by 2005; in the other four countries attempts to introduce fluoridated salt were not successful.

History

1909 photograph by Frederick McKay of G.V. Black (left), Isaac Burton and F.Y. Wilson, studying the Colorado brown stain.

The history of water fluoridation can be divided into three periods. The first (c. 1801–1933) was research into the cause of a form of mottled tooth enamel called the Colorado brown stain. The second (c. 1933–1945) focused on the relationship between fluoride concentrations, fluorosis, and tooth decay, and established that moderate levels of fluoride prevent cavities. The third period, from 1945 on, focused on adding fluoride to community water supplies.

In the first half of the 19th century, investigators established that fluoride occurs with varying concentrations in teeth, bone, and drinking water. In the second half they speculated that fluoride would protect against tooth decay, proposed supplementing the diet with fluoride, and observed mottled enamel (now called severe dental fluorosis) without knowing the cause. In 1874, the German public health officer Carl Wilhelm Eugen Erhardt recommended potassium fluoride supplements to preserve teeth. In 1892 the British physician James Crichton-Browne noted in an address that fluoride's absence from diets had resulted in teeth that were "peculiarly liable to decay", and who proposed "the reintroduction into our diet of fluorine in some suitable natural form ... to fortify the teeth of the next generation".

The foundation of water fluoridation in the U.S. was the research of the dentist Frederick McKay (1874–1959). McKay spent thirty years investigating the cause of what was then known as the Colorado brown stain, which produced mottled but also cavity-free teeth; with the help of G.V. Black and other researchers, he established that the cause was fluoride. The first report of a statistical association between the stain and lack of tooth decay was made by UK dentist Norman Ainsworth in 1925. In 1931, an Alcoa chemist, H.V. Churchill, concerned about a possible link between aluminum and staining, analyzed water from several areas where the staining was common and found that fluoride was the common factor.

H. Trendley Dean set out in 1931 to study fluoride's harm,
but by 1950 had demonstrated the cavity-prevention effects of small amounts.

In the 1930s and early 1940s, H. Trendley Dean and colleagues at the newly created U.S. National Institutes of Health published several epidemiological studies suggesting that a fluoride concentration of about 1 mg/L was associated with substantially fewer cavities in temperate climates, and that it increased fluorosis but only to a level that was of no medical or aesthetic concern. Other studies found no other significant adverse effects even in areas with fluoride levels as high as 8 mg/L. To test the hypothesis that adding fluoride would prevent cavities, Dean and his colleagues conducted a controlled experiment by fluoridating the water in Grand Rapids, Michigan, starting January 25, 1945. The results, published in 1950, showed significant reduction of cavities. Significant reductions in tooth decay were also reported by important early studies outside the U.S., including the Brantford–Sarnia–Stratford study in Canada (1945–1962), the Tiel–Culemborg study in the Netherlands (1953–1969), the Hastings study in New Zealand (1954–1970), and the Department of Health study in the U.K. (1955–1960). By present-day standards these and other pioneering studies were crude, but the large reductions in cavities convinced public health professionals of the benefits of fluoridation.

Fluoridation became an official policy of the U.S. Public Health Service by 1951, and by 1960 water fluoridation had become widely used in the U.S., reaching about 50 million people. By 2006, 69.2% of the U.S. population on public water systems were receiving fluoridated water, amounting to 61.5% of the total U.S. population; 3.0% of the population on public water systems were receiving naturally occurring fluoride. In some other countries the pattern was similar. New Zealand, which led the world in per-capita sugar consumption and had the world's worst teeth, began fluoridation in 1953, and by 1968 fluoridation was used by 65% of the population served by a piped water supply. Fluoridation was introduced into Brazil in 1953, was regulated by federal law starting in 1974, and by 2004 was used by 71% of the population. In the Republic of Ireland, fluoridation was legislated in 1960, and after a constitutional challenge the two major cities of Dublin and Cork began it in 1964; fluoridation became required for all sizeable public water systems and by 1996 reached 66% of the population. In other locations, fluoridation was used and then discontinued: in Kuopio, Finland, fluoridation was used for decades but was discontinued because the school dental service provided significant fluoride programs and the cavity risk was low, and in Basel, Switzerland, it was replaced with fluoridated salt.

McKay's work had established that fluorosis occurred before tooth eruption. Dean and his colleagues assumed that fluoride's protection against cavities was also pre-eruptive, and this incorrect assumption was accepted for years. By 2000, however, the topical effects of fluoride (in both water and toothpaste) were well understood, and it had become known that a constant low level of fluoride in the mouth works best to prevent cavities.

Economics

Fluoridation costs an estimated $1.04 per person-year on the average (range: $0.24–$10.96; all costs in this paragraph are for the U.S. and are in 2016 dollars, inflation-adjusted from earlier estimates). Larger water systems have lower per capita cost, and the cost is also affected by the number of fluoride injection points in the water system, the type of feeder and monitoring equipment, the fluoride chemical and its transportation and storage, and water plant personnel expertise. In affluent countries the cost of salt fluoridation is also negligible; developing countries may find it prohibitively expensive to import the fluoride additive. By comparison, fluoride toothpaste costs an estimated $9–$17 per person-year, with the incremental cost being zero for people who already brush their teeth for other reasons; and dental cleaning and application of fluoride varnish or gel costs an estimated $95 per person-year. Assuming the worst case, with the lowest estimated effectiveness and highest estimated operating costs for small cities, fluoridation costs an estimated $16–$24 per saved tooth-decay surface, which is lower than the estimated $93 to restore the surface and the estimated $158 average discounted lifetime cost of the decayed surface, which includes the cost to maintain the restored tooth surface. It is not known how much is spent in industrial countries to treat dental fluorosis, which is mostly due to fluoride from swallowed toothpaste.

Although a 1989 workshop on cost-effectiveness of cavity prevention concluded that water fluoridation is one of the few public health measures that save more money than they cost, little high-quality research has been done on the cost-effectiveness and solid data are scarce. Dental sealants are cost-effective only when applied to high-risk children and teeth. A 2002 U.S. review estimated that on average, sealing first permanent molars saves costs when they are decaying faster than 0.47 surfaces per person-year whereas water fluoridation saves costs when total decay incidence exceeds 0.06 surfaces per person-year. In the U.S., water fluoridation is more cost-effective than other methods to reduce tooth decay in children, and a 2008 review concluded that water fluoridation is the best tool for combating cavities in many countries, particularly among socially disadvantaged groups. A 2016 review of studies published between 1995 to 2013 found that water fluoridation in the U.S. was cost-effective, and that it was more so in larger communities.

U.S. data from 1974 to 1992 indicate that when water fluoridation is introduced into a community, there are significant decreases in the number of employees per dental firm and the number of dental firms. The data suggest that some dentists respond to the demand shock by moving to non-fluoridated areas and by retraining as specialists.

Controversy

The water fluoridation controversy arises from political, moral, ethical, economic, and safety concerns regarding the fluoridation of public water supplies. Those opposed argue that water fluoridation may cause serious health problems, is not effective enough to justify the costs, and has a dosage that cannot be precisely controlled.

With regard to ethics, like vaccination and food fortification, fluoridation pits the common good against individual rights.

Opposition to fluoridation has existed since its initiation in the 1940s. During the 1950s and 1960s, conspiracy theorists claimed that fluoridation was a communist plot to undermine American public health. In recent years water fluoridation has become a pervasive health and political issue in many countries, resulting in changes to public policy regarding water fluoridation.

Public health authorities find a medical consensus that water fluoridation at appropriate levels is a safe and effective means to prevent suffering and promote oral health, and generally support fluoridation. Opponents of fluoridation include some researchers, dental and medical professionals, alternative medical practitioners, health food enthusiasts, a few religious groups (mostly Christian Scientists in the U.S.), and occasionally consumer groups and environmentalists. Organized political opposition has come from libertarians, the John Birch Society, and from groups like the Green parties in the UK and New Zealand.

Water Purification

Control room and schematics of the water purification plant to Lac de Bret, Switzerland

Water purification is the process of removing undesirable chemicals, biological contaminants, suspended solids and gases from water. The goal is to produce water fit for a specific purpose. Most water is disinfected for human consumption (drinking water), but water purification may also be designed for a variety of other purposes, including fulfilling the requirements of medical, pharmacological, chemical and industrial applications. The methods used include physical processes such as filtration, sedimentation, and distillation; biological processes such as slow sand filters or biologically active carbon; chemical processes such as flocculation and chlorination and the use of electromagnetic radiation such as ultraviolet light.

Purifying water may reduce the concentration of particulate matter including suspended particles, parasites, bacteria, algae, viruses, fungi, as well as reducing the amount of a range of dissolved and particulate material derived from the surfaces that come from runoff due to rain.

The standards for drinking water quality are typically set by governments or by international standards. These standards usually include minimum and maximum concentrations of contaminants, depending on the intended purpose of water use.

Visual inspection cannot determine if water is of appropriate quality. Simple procedures such as boiling or the use of a household activated carbon filter are not sufficient for treating all the possible contaminants that may be present in water from an unknown source. Even natural spring water – considered safe for all practical purposes in the 19th century – must now be tested before determining what kind of treatment, if any, is needed. Chemical and microbiological analysis, while expensive, are the only way to obtain the information necessary for deciding on the appropriate method of purification.

According to a 2007 World Health Organization (WHO) report, 1.1 billion people lack access to an improved drinking water supply, 88% of the 4 billion annual cases of diarrheal disease are attributed to unsafe water and inadequate sanitation and hygiene, while 1.8 million people die from diarrheal diseases each year. The WHO estimates that 94% of these diarrheal cases are preventable through modifications to the environment, including access to safe water. Simple techniques for treating water at home, such as chlorination, filters, and solar disinfection, and storing it in safe containers could save a huge number of lives each year. Reducing deaths from waterborne diseases is a major public health goal in developing countries.

Sources of Water

- Groundwater: The water emerging from some deep ground water may have fallen as rain many tens, hundreds, or thousands of years ago. Soil and rock layers naturally filter the ground water to a high degree of clarity and often, it does not require additional treatment besides adding chlorine or chloramines as secondary disinfectants. Such water may emerge as springs, artesian springs, or may be extracted from boreholes or wells. Deep ground water is generally of very high bacteriological quality (i.e., pathogenic bacteria or the pathogenic protozoa are typically absent), but the water may be rich in dissolved solids, especially carbonates and sulfates of calcium and magnesium. Depending on the strata through which the water has flowed, other ions may also be present including chloride, and bicarbonate. There may be a requirement to reduce the iron or manganese content of this water to make it acceptable for drinking, cooking, and laundry use. Primary disinfection may also be required. Where groundwater recharge is practiced (a process in which river water is injected into an aquifer to store the water in times of plenty so that it is available in times of drought), the groundwater may require additional treatment depending on applicable state and federal regulations.
- Upland lakes and reservoirs: Typically located in the headwaters of river systems, upland reservoirs are usually sited above any human habitation and may be surrounded by a protective zone to restrict the opportunities for contamination. Bacteria and pathogen levels are usually low, but some bacteria, protozoa or algae will be present. Where uplands are forested or peaty, humic acids can colour the water. Many upland sources have low pH which require adjustment.

1. Rivers, canals and low land reservoirs: Low land surface waters will have a significant bacterial load and may also contain algae, suspended solids and a variety of dissolved constituents.
2. Atmospheric water generation is a new technology that can provide high quality drinking water by extracting water from the air by cooling the air and thus condensing water vapor.
3. Rainwater harvesting or fog collection which collect water from the atmosphere can be used especially in areas with significant dry seasons and in areas which experience fog even when there is little rain.
4. Desalination of seawater by distillation or reverse osmosis.
5. Surface Water: Freshwater bodies that are open to the atmosphere and are not designated as groundwater are termed surface waters.

Treatment

Aims

The aims of the treatment are to remove unwanted constituents in the water and to make it safe to drink or fit for a specific purpose in industry or medical applications. Widely varied techniques are available to remove contaminants like fine solids, micro-organisms and some dissolved inorganic and organic materials, or environmental persistent pharmaceutical pollutants. The choice of method will depend on the quality of the water being treated, the cost of the treatment process and the quality standards expected of the processed water.

The processes below are the ones commonly used in water purification plants. Some or most may not be used depending on the scale of the plant and quality of the raw (source) water.

Pretreatment

1. Pumping and containment – The majority of water must be pumped from its source or directed into pipes or holding tanks. To avoid adding contaminants to the water, this physical infrastructure must be made from appropriate materials and constructed so that accidental contamination does not occur.
2. Screening – The first step in purifying surface water is to remove large debris such as sticks, leaves, rubbish and other large particles which may interfere with subsequent purification steps. Most deep groundwater does not need screening before other purification steps.
3. Storage – Water from rivers may also be stored in bankside reservoirs for pe-

riods between a few days and many months to allow natural biological purification to take place. This is especially important if treatment is by slow sand filters. Storage reservoirs also provide a buffer against short periods of drought or to allow water supply to be maintained during transitory pollution incidents in the source river.

4. Pre-chlorination – In many plants the incoming water was chlorinated to minimize the growth of fouling organisms on the pipe-work and tanks. Because of the potential adverse quality effects, this has largely been discontinued.

pH Adjustment

Pure water has a pH close to 7 (neither alkaline nor acidic). Sea water can have pH values that range from 7.5 to 8.4 (moderately alkaline). Fresh water can have widely ranging pH values depending on the geology of the drainage basin or aquifer and the influence of contaminant inputs (acid rain). If the water is acidic (lower than 7), lime, soda ash, or sodium hydroxide can be added to raise the pH during water purification processes. Lime addition increases the calcium ion concentration, thus raising the water hardness. For highly acidic waters, forced draft degasifiers can be an effective way to raise the pH, by stripping dissolved carbon dioxide from the water. Making the water alkaline helps coagulation and flocculation processes work effectively and also helps to minimize the risk of lead being dissolved from lead pipes and from lead solder in pipe fittings. Sufficient alkalinity also reduces the corrosiveness of water to iron pipes. Acid (carbonic acid, hydrochloric acid or sulfuric acid) may be added to alkaline waters in some circumstances to lower the pH. Alkaline water (above pH 7.0) does not necessarily mean that lead or copper from the plumbing system will not be dissolved into the water. The ability of water to precipitate calcium carbonate to protect metal surfaces and reduce the likelihood of toxic metals being dissolved in water is a function of pH, mineral content, temperature, alkalinity and calcium concentration.

Coagulation and Flocculation

One of the first steps in a conventional water purification process is the addition of chemicals to assist in the removal of particles suspended in water. Particles can be inorganic such as clay and silt or organic such as algae, bacteria, viruses, protozoa and natural organic matter. Inorganic and organic particles contribute to the turbidity and color of water.

The addition of inorganic coagulants such as aluminum sulfate (or alum) or iron (III) salts such as iron(III) chloride cause several simultaneous chemical and physical interactions on and among the particles. Within seconds, negative charges on the particles are neutralized by inorganic coagulants. Also within seconds, metal hydroxide precipitates of the iron and aluminium ions begin to form. These precipitates combine into larger particles under natural processes such as Brownian motion and through induced mixing which is sometimes referred to as flocculation. The term most often used for the amorphous metal hydroxides is “floc.” Large, amorphous aluminum and iron (III) hydroxides adsorb and enmesh particles in suspension and facilitate the removal of particles by subsequent processes of sedimentation and filtration.

Aluminum hydroxides are formed within a fairly narrow pH range, typically: 5.5 to about 7.7. Iron (III) hydroxides can form over a larger pH range including pH levels lower than are effective for alum, typically: 5.0 to 8.5.

In the literature, there is much debate and confusion over the usage of the terms coagulation and flocculation—where does coagulation end and flocculation begin? In water purification plants, there is usually a high energy, rapid mix unit process (detention time in seconds) where the coagulant chemicals are added followed by flocculation basins (detention times range from 15 to 45 minutes) where low energy inputs turn large paddles or other gentle mixing devices to enhance the formation of floc. In fact, coagulation and flocculation processes are ongoing once the metal salt coagulants are added.

Organic polymers were developed in the 1960s as aids to coagulants and, in some cases, as replacements for the inorganic metal salt coagulants. Synthetic organic polymers are high molecular weight compounds that carry negative, positive or neutral charges. When organic polymers are added to water with particulates, the high molecular weight compounds adsorb onto particle surfaces and through interparticle bridging coalesce with other particles to form floc. PolyDADMAC is a popular cationic (positively charged) organic polymer used in water purification plants.

Sedimentation

Waters exiting the flocculation basin may enter the sedimentation basin, also called a clarifier or settling basin. It is a large tank with low water velocities, allowing floc to settle to the bottom. The sedimentation basin is best located close to the flocculation basin so the transit between the two processes does not permit settlement or floc break up. Sedimentation basins may be rectangular, where water flows from end to end, or circular where flow is from the centre outward. Sedimentation basin outflow is typically over a weir so only a thin top layer of water—that furthest from the sludge—exits.

In 1904, Allen Hazen showed that the efficiency of a sedimentation process was a function of the particle settling velocity, the flow through the tank and the surface area of tank. Sedimentation tanks are typically designed within a range of overflow rates of 0.5 to 1.0 gallons per minute per square foot (or 1.25 to 2.5 meters per hour). In general, sedimentation basin efficiency is not a function of detention time or depth of the basin. Although, basin depth must be sufficient so that water currents do not disturb the sludge and settled particle interactions are promoted. As particle concentrations in the settled water increase near the sludge surface on the bottom of the tank, settling velocities can increase due to collisions and agglomeration of particles. Typical detention times for sedimentation vary from 1.5 to 4 hours and basin depths vary from 10 to 15 feet (3 to 4.5 meters).

Inclined flat plates or tubes can be added to traditional sedimentation basins to improve particle removal performance. Inclined plates and tubes drastically increase the surface area available for particles to be removed in concert with Hazen's original theory. The amount of ground surface area occupied by a sedimentation basin with inclined plates or tubes can be far smaller than a conventional sedimentation basin.

Sludge Storage and Removal

As particles settle to the bottom of a sedimentation basin, a layer of sludge is formed on the floor of the tank which must be removed and treated. The amount of sludge generated is significant, often 3 to 5 percent of the total volume of water to be treated. The cost of treating and disposing of the sludge can impact the operating cost of a water treatment plant. The sedimentation basin may be equipped with mechanical cleaning devices that continually clean its bottom, or the basin can be periodically taken out of service and cleaned manually.

Floc Blanket Clarifiers

A subcategory of sedimentation is the removal of particulates by entrapment in a layer of suspended floc as the water is forced upward. The major advantage of floc blanket clarifiers is that they occupy a smaller footprint than conventional sedimentation. Disadvantages are that particle removal efficiency can be highly variable depending on changes in influent water quality and influent water flow rate.

Dissolved Air Flotation

When particles to be removed do not settle out of solution easily, dissolved air flotation (DAF) is often used. After coagulation and flocculation processes, water flows to DAF tanks where air diffusers on the tank bottom create fine bubbles that attach to floc resulting in a floating mass of concentrated floc. The floating floc blanket is removed from the surface and clarified water is withdrawn from the bottom of the DAF tank. Water supplies that are particularly vulnerable to unicellular algae blooms and supplies with low turbidity and high colour often employ DAF.

Filtration

After separating most floc, the water is filtered as the final step to remove remaining suspended particles and unsettled floc.

Rapid sand Filters

The most common type of filter is a rapid sand filter. Water moves vertically through sand which often has a layer of activated carbon or anthracite coal above the sand. The top layer removes organic compounds, which contribute to taste and odour. The space between sand particles is larger than the smallest suspended particles, so simple filtration is not enough. Most particles pass through surface layers but are trapped in pore spaces or adhere to sand particles. Effective filtration extends into the depth of the filter. This property of the filter is key to its operation: if the top layer of sand were to block all the particles, the filter would quickly clog.

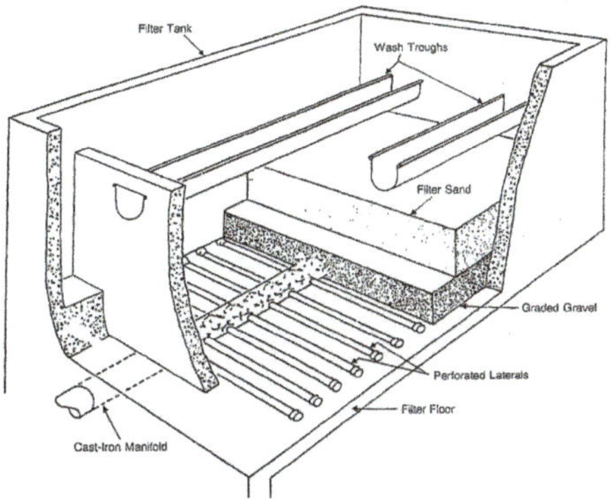

Cutaway view of a typical rapid sand filter

To clean the filter, water is passed quickly upward through the filter, opposite the normal direction (called *backflushing* or *backwashing*) to remove embedded or unwanted particles. Prior to this step, compressed air may be blown up through the bottom of the filter to break up the compacted filter media to aid the backwashing process; this is known as *air scouring*. This contaminated water can be disposed of, along with the sludge from the sedimentation basin, or it can be recycled by mixing with the raw water entering the plant although this is often considered poor practice since it re-introduces an elevated concentration of bacteria into the raw water.

Some water treatment plants employ pressure filters. These work on the same principle as rapid gravity filters, differing in that the filter medium is enclosed in a steel vessel and the water is forced through it under pressure.

Advantages:

- Filters out much smaller particles than paper and sand filters can.
- Filters out virtually all particles larger than their specified pore sizes.
- They are quite thin and so liquids flow through them fairly rapidly.
- They are reasonably strong and so can withstand pressure differences across them of typically 2–5 atmospheres.
- They can be cleaned (back flushed) and reused.

Slow Sand Filters

Slow sand filters may be used where there is sufficient land and space, as the water must be passed very slowly through the filters. These filters rely on biological treatment processes for their action rather than physical filtration. The filters are carefully constructed using graded layers of sand, with the coarsest sand, along with some gravel, at the bottom and finest sand at the top. Drains at the base convey treated water away for disinfection. Filtration depends on the development of a thin biological layer, called the zoogleal layer or Schmutzdecke, on the surface of the filter. An effective slow sand filter may remain in service for many weeks or even months if the pretreatment is well designed and produces water with a very low available nutrient level which physical methods of treatment rarely achieve. Very low nutrient levels allow water to be safely sent through distribution systems with very low disinfectant levels, thereby reducing consumer irritation over offensive levels of chlorine and chlorine by-products. Slow sand filters are not backwashed; they are maintained by having the top layer of sand scraped off when flow is eventually obstructed by biological growth.

Slow "artificial" filtration (a variation of bank filtration) to the ground, Water purification plant Káraný, Czech Republic.

A profile of layers of gravel, sand and fine sand used in a slow sand filter plant.

A specific "large-scale" form of slow sand filter is the process of bank filtration, in which natural sediments in a riverbank are used to provide a first stage of contaminant filtration. While typically not clean enough to be used directly for drinking water, the water gained from the associated extraction wells is much less problematic than river water taken directly from the major streams where bank filtration is often used.

Membrane Filtration

Membrane filters are widely used for filtering both drinking water and sewage. For drinking water, membrane filters can remove virtually all particles larger than 0.2 µm—including *giardia* and *cryptosporidium*. Membrane filters are an effective form of tertiary treatment when it is desired to reuse the water for industry, for limited domestic purposes, or before discharging the water into a river that is used by towns further downstream. They are widely used in industry, particularly for beverage preparation (including bottled water). However no filtration can remove substances that are actually dissolved in the water such as phosphorus, nitrates and heavy metal ions.

Removal of Ions and other Dissolved Substances

Ultrafiltration membranes use polymer membranes with chemically formed microscopic pores that can be used to filter out dissolved substances avoiding the use of coagulants. The type of membrane media determines how much pressure is needed to drive the water through and what sizes of micro-organisms can be filtered out.

Ion exchange: Ion exchange systems use ion exchange resin- or zeolite-packed columns to replace unwanted ions. The most common case is water softening consisting of removal of Ca^{2+} and Mg^{2+} ions replacing them with benign (soap friendly) Na^+ or K^+ ions. Ion exchange resins are also used to remove toxic ions such as nitrite, lead, mercury, arsenic and many others.

Precipitative softening: Water rich in hardness (calcium and magnesium ions) is treated with lime (calcium oxide) and/or soda-ash (sodium carbonate) to precipitate calcium carbonate out of solution utilizing the common-ion effect.

Electrodeionization: Water is passed between a positive electrode and a negative electrode. Ion exchange membranes allow only positive ions to migrate from the treated water toward the negative electrode and only negative ions toward the positive electrode. High purity deionized water is produced continuously, similar to ion exchange treatment. Complete removal of ions from water is possible if the right conditions are met. The water is normally pre-treated with a reverse osmosis unit to remove non-ionic organic contaminants, and with gas transfer membranes to remove carbon dioxide. A water recovery of 99% is possible if the concentrate stream is fed to the RO inlet.

Disinfection

Disinfection is accomplished both by filtering out harmful micro-organisms and also by adding disinfectant chemicals. Water is disinfected to kill any pathogens which pass through the filters and to provide a residual dose of disinfectant to kill or inactivate potentially harmful micro-organisms in the storage and distribution systems. Possible pathogens include viruses, bacteria, including *Salmonella*, *Cholera*, *Campylobacter* and *Shigella*, and protozoa, including *Giardia lamblia* and other *cryptosporidia*. Following the introduction of any chemical disinfecting agent, the water is usually held in temporary storage – often called a contact tank or clear well to allow the disinfecting action to complete.

Pumps used to add required amount of chemicals to the clear water at the water purification plant before the distribution. From left to right: sodium hypochlorite for disinfection, zinc orthophosphate as a corrosion inhibitor, sodium hydroxide for pH adjustment, and fluoride for tooth decay prevention.

Chlorine Disinfection

The most common disinfection method involves some form of chlorine or its compounds such as chloramine or chlorine dioxide. Chlorine is a strong oxidant that rapidly kills many harmful micro-organisms. Because chlorine is a toxic gas, there is a danger of a release associated with its use. This problem is avoided by the use of sodium hypochlorite, which is a relatively inexpensive solution used in household bleach that releases free chlorine when dissolved in water. Chlorine solutions can be generated on site by electrolyzing common salt solutions. A solid form, calcium hypochlorite, releases chlorine on contact with water. Handling the solid, however, requires greater routine human contact through opening bags and pouring than the use of gas cylinders or bleach which are more easily automated. The generation of liquid sodium hypochlorite is both inexpensive and safer than the use of gas or solid chlorine.

All forms of chlorine are widely used, despite their respective drawbacks. One drawback is that chlorine from any source reacts with natural organic compounds in the water to form potentially harmful chemical by-products. These by-products, trihalomethanes (THMs) and haloacetic acids (HAAs), are both carcinogenic in large quantities and are regulated by the United States Environmental Protection Agency (EPA) and the Drinking Water Inspectorate in the UK. The formation of THMs and haloacetic acids may be minimized by effective removal of as many organics from the water as possible prior to chlorine addition. Although chlorine is effective in killing bacteria, it has limited effectiveness against protozoa that form cysts in water (*Giardia lamblia* and *Cryptosporidium*, both of which are pathogenic).

Chlorine Dioxide Disinfection

Chlorine dioxide is a faster-acting disinfectant than elemental chlorine. It is relatively rarely used, because in some circumstances it may create excessive amounts of chlorite, which is a by-product regulated to low allowable levels in the United States. Chlorine dioxide can be supplied as an aqueous solution and added to water to avoid gas handling problems; chlorine dioxide gas accumulations may spontaneously detonate.

Chloramine Disinfection

The use of chloramine is becoming more common as a disinfectant. Although chloramine is not as strong an oxidant, it does provide a longer-lasting residual than free chlorine and it will not readily form THMs or haloacetic acids. It is possible to convert chlorine to chloramine by adding ammonia to the water after addition of chlorine. The chlorine and ammonia react to form chloramine. Water distribution systems disinfected with chloramines may experience nitrification, as ammonia is a nutrient for bacterial growth, with nitrates being generated as a by-product.

Ozone Disinfection

Ozone is an unstable molecule which readily gives up one atom of oxygen providing a powerful oxidizing agent which is toxic to most waterborne organisms. It is a very strong, broad spectrum disinfectant that is widely used in Europe. It is an effective method to inactivate harmful protozoa that form cysts. It also works well against almost all other pathogens. Ozone is made by passing oxygen through ultraviolet light or a "cold" electrical discharge. To use ozone as a disinfectant, it must be created on-site and added to the water by bubble contact. Some of the advantages of ozone include the production of fewer dangerous by-products and the absence of taste and odour problems (in comparison to chlorination). Another advantage of ozone is that it leaves no residual disinfectant in the water. Ozone has been used in drinking water plants since 1906 where the first industrial ozonation plant was built in Nice, France. The U.S. Food and Drug Administration has accepted ozone as being safe; and it is applied as an anti-microbiological agent for the treatment, storage, and processing of foods. However, although fewer by-products are formed by ozonation, it has been discovered that ozone reacts with bromide ions in water to produce concentrations of the suspected carcinogen bromate. Bromide can be found in fresh water supplies in sufficient concentrations to produce (after ozonation) more than 10 parts per billion (ppb) of bromate — the maximum contaminant level established by the USEPA.

Ultraviolet Disinfection

Ultraviolet light (UV) is very effective at inactivating cysts, in low turbidity water. UV light's disinfection effectiveness decreases as turbidity increases, a result of the absorption, scattering, and shadowing caused by the suspended solids. The main disadvantage to the use of UV radiation is that, like ozone treatment, it leaves no residual disinfectant in the water; therefore, it is sometimes necessary to add a residual disinfectant after the primary disinfection process. This is often done through the addition of chloramines, discussed above as a primary disinfectant. When used in this manner, chloramines provide an effective residual disinfectant with very few of the negative effects of chlorination.

Portable Water Purification

Portable water purification devices and methods are available for disinfection and treatment in emergencies or in remote locations. Disinfection is the primary goal, since aesthetic considerations such as taste, odor, appearance, and trace chemical contamination do not affect the short-term safety of drinking water.

Additional Treatment Options

- Water fluoridation: in many areas fluoride is added to water with the goal of preventing tooth decay. Fluoride is usually added after the disinfection process. In the U.S., fluoridation is usually accomplished by the addition of hexafluorosilicic acid, which decomposes in water, yielding fluoride ions.
- Water conditioning: This is a method of reducing the effects of hard water. In water systems subject to heating hardness salts can be deposited as the decomposition of bicarbonate ions creates carbonate ions that precipitate out of solution. Water with high concentrations of hardness salts can be treated with soda ash (sodium carbonate) which precipitates out the excess salts, through the common-ion effect, producing calcium carbonate of very high purity. The precipitated calcium carbonate is traditionally sold to the manufacturers of toothpaste. Several other methods of industrial and residential water treatment are claimed (without general scientific acceptance) to include the use of magnetic and/or electrical fields reducing the effects of hard water.
- Plumbosolvency reduction: In areas with naturally acidic waters of low conductivity (i.e. surface rainfall in upland mountains of igneous rocks), the water may be capable of dissolving lead from any lead pipes that it is carried in. The addition of small quantities of phosphate ion and increasing the pH slightly both assist in greatly reducing plumbo-solvency by creating insoluble lead salts on the inner surfaces of the pipes.
- Radium Removal: Some groundwater sources contain radium, a radioactive chemical element. Typical sources include many groundwater sources north of

the Illinois River in Illinois. Radium can be removed by ion exchange, or by water conditioning. The back flush or sludge that is produced is, however, a low-level radioactive waste.

- Fluoride Removal: Although fluoride is added to water in many areas, some areas of the world have excessive levels of natural fluoride in the source water. Excessive levels can be toxic or cause undesirable cosmetic effects such as staining of teeth. Methods of reducing fluoride levels is through treatment with activated alumina and bone char filter media.

Other Water Purification Techniques

Other popular methods for purifying water, especially for local private supplies are listed below. In some countries some of these methods are also used for large scale municipal supplies. Particularly important are distillation (de-salination of seawater) and reverse osmosis.

1. Boiling: Bringing it to its boiling point at 100 °C (212 °F), is the oldest and most effective way since it eliminates most microbes causing intestine related diseases, but it cannot remove chemical toxins or impurities. For human health, complete sterilization of water is not required, since the heat resistant microbes are not intestine affecting. The traditional advice of boiling water for ten minutes is mainly for additional safety, since microbes start getting eliminated at temperatures greater than 60 °C (140 °F). Though the boiling point decreases with increasing altitude, it is not enough to affect the disinfecting process. In areas where the water is "hard" (that is, containing significant dissolved calcium salts), boiling decomposes the bicarbonate ions, resulting in partial precipitation as calcium carbonate. This is the "fur" that builds up on kettle elements, etc., in hard water areas. With the exception of calcium, boiling does not remove solutes of higher boiling point than water and in fact increases their concentration (due to some water being lost as vapour). Boiling does not leave a residual disinfectant in the water. Therefore, water that is boiled and then stored for any length of time may acquire new pathogens.

2. Granular Activated Carbon filtering: a form of activated carbon with a high surface area, adsorbs many compounds including many toxic compounds. Water passing through activated carbon is commonly used in municipal regions with organic contamination, taste or odors. Many household water filters and fish tanks use activated carbon filters to further purify the water. Household filters for drinking water sometimes contain silver as metallic silver nanoparticle. If water is held in the carbon block for longer periods, microorganisms can grow inside which results in fouling and contamination. Silver nanoparticles are excellent anti-bacterial material and they can decompose toxic halo-organic compounds such as pesticides into non-toxic organic products.

3. Distillation involves boiling the water to produce water vapour. The vapour contacts a cool surface where it condenses as a liquid. Because the solutes are not normally vaporised, they remain in the boiling solution. Even distillation does not completely purify water, because of contaminants with similar boiling points and droplets of unvapourised liquid carried with the steam. However, 99.9% pure water can be obtained by distillation.
4. Reverse osmosis: Mechanical pressure is applied to an impure solution to force pure water through a semi-permeable membrane. Reverse osmosis is theoretically the most thorough method of large scale water purification available, although perfect semi-permeable membranes are difficult to create. Unless membranes are well-maintained, algae and other life forms can colonize the membranes.
5. The use of iron in removing arsenic from water.
6. Direct contact membrane distillation (DCMD). Applicable to desalination. Heated seawater is passed along the surface of a hydrophobic polymer membrane. Evaporated water passes from the hot side through pores in the membrane into a stream of cold pure water on the other side. The difference in vapour pressure between the hot and cold side helps to push water molecules through.
7. Desalination – is a process by which saline water (generally sea water) is converted to fresh water. The most common desalination processes are distillation and reverse osmosis. Desalination is currently expensive compared to most alternative sources of water, and only a very small fraction of total human use is satisfied by desalination. It is only economically practical for high-valued uses (such as household and industrial uses) in arid areas.
8. Gas hydrate crystals centrifuge method. If carbon dioxide or other low molecular weight gas is mixed with contaminated water at high pressure and low temperature, gas hydrate crystals will form exothermically. Separation of the crystalline hydrate may be performed by centrifuge or sedimentation and decanting. Water can be released from the hydrate crystals by heating
9. In Situ Chemical Oxidation, a form of advanced oxidation processes and advanced oxidation technology, is an environmental remediation technique used for soil and/or groundwater remediation to reduce the concentrations of targeted environmental contaminants to acceptable levels. ISCO is accomplished by injecting or otherwise introducing strong chemical oxidizers directly into the contaminated medium (soil or groundwater) to destroy chemical contaminants in place. It can be used to remediate a variety of organic compounds, including some that are resistant to natural degradation
10. Bioremediation is a technique that uses microorganisms in order to remove or extract certain waste products from a contaminated area. Since 1991 bioreme-

diation has been a suggested tactic to remove impurities from water such as alkanes, perchlorates, and metals. The treatment of ground and surface water, through bioremediation, with respect to perchlorate and chloride compounds, has seen success as perchlorate compounds are highly soluble making it difficult to remove. Such success by use of *Dechloromonas agitata* strain CKB include field studies conducted in Maryland and the Southwest region of the United States. Although a bioremediation technique may be successful, implementation is not feasible as there is still much to be studied regarding rates and after effects of microbial activity as well as producing a large scale implementation method.

Safety and Controversies

Drinking water pollution detector Rainbow trout (*Oncorhynchus mykiss*) are being used in water purification plants to detect acute water pollution.

In April, 2007, the water supply of Spencer, Massachusetts became contaminated with excess sodium hydroxide (lye) when its treatment equipment malfunctioned.

Many municipalities have moved from free chlorine to chloramine as a disinfection agent. However, chloramine appears to be a corrosive agent in some water systems. Chloramine can dissolve the "protective" film inside older service lines, leading to the leaching of lead into residential spigots. This can result in harmful exposure, including elevated blood lead levels. Lead is a known neurotoxin.

Demineralized Water

Distillation removes all minerals from water, and the membrane methods of reverse osmosis and nanofiltration remove most to all minerals. This results in demineralized water which is not considered ideal drinking water. The World Health Organization has investigated the health effects of demineralized water since 1980. Experiments in humans found that demineralized water increased diuresis and the elimination of electrolytes, with decreased blood serum potassium concentration. Magnesium, calcium, and other minerals in water can help to protect against nutritional deficiency. Demineralized water may also increase the risk from toxic metals because it more readily leaches materials from piping like lead and cadmium, which is prevented by dissolved minerals such as calcium and magnesium. Low-mineral water has been implicated in specific cases of lead poisoning in infants, when lead from pipes leached at especially high rates into the water. Recommendations for magnesium have been put at a minimum of 10 mg/L with 20–30 mg/L optimum; for calcium a 20 mg/L minimum and a 40–80 mg/L optimum, and a total water hardness (adding magnesium and calcium) of 2 to 4 mmol/L. At water hardness above 5 mmol/L, higher incidence of gallstones, kidney stones, urinary stones, arthrosis, and arthropathies have been observed. Additionally, desalination processes can increase the risk of bacterial contamination.

Manufacturers of home water distillers claim the opposite—that minerals in water are the cause of many diseases, and that most beneficial minerals come from food, not water. They quote the American Medical Association as saying "The body's need for minerals is largely met through foods, not drinking water." The WHO report agrees that "drinking water, with some rare exceptions, is not the major source of essential elements for humans" and is "not the major source of our calcium and magnesium intake", yet states that demineralized water is harmful anyway. "Additional evidence comes from animal experiments and clinical observations in several countries. Animals given zinc or magnesium dosed in their drinking water had a significantly higher concentration of these elements in the serum than animals given the same elements in much higher amounts with food and provided with low-mineral water to drink."

Demineralized water is also the type of water injection used to increase thrust in some turbojet engines such as the J57 engine used on KC-135A/Q and B-52A/B/C/D/E/F/G aircraft, as well as the civilian version of that engine, the JT-3C used on early build Boeing 707 and Douglas DC-8 airliners.

History

Drawing of an apparatus for studying the chemical analysis of mineral waters in a book from 1799.

The first experiments into water filtration were made in the 17th century. Sir Francis Bacon attempted to desalinate sea water by passing the flow through a sand filter. Although his experiment did not succeed, it marked the beginning of a new interest in the field. The fathers of microscopy, Antonie van Leeuwenhoek and Robert Hooke, used the newly invented microscope to observe for the first time small material particles that lay suspended in the water, laying the groundwork for the future understanding of waterborne pathogens.

Sand Filter

The first documented use of sand filters to purify the water supply dates to 1804, when the owner of a bleachery in Paisley, Scotland, John Gibb, installed an experimental filter, selling his unwanted surplus to the public. This method was refined in the following two decades by engineers working for private water companies, and it culminated in the first treated public water supply in the world, installed by engineer James Simpson for the Chelsea Waterworks Company in London in 1829. This installation provided filtered water for every resident of the area, and the network design was widely copied throughout the United Kingdom in the ensuing decades.

The practice of water treatment soon became mainstream and common, and the virtues of the system were made starkly apparent after the investigations of the physician John Snow during the 1854 Broad Street cholera outbreak. Snow was sceptical of the then-dominant miasma theory that stated that diseases were caused by noxious "bad airs". Although the germ theory of disease had not yet been developed, Snow's observations led him to discount the prevailing theory. His 1855 essay *On the Mode of Communication of Cholera* conclusively demonstrated the role of the water supply in spreading the cholera epidemic in Soho, with the use of a dot distribution map and statistical proof to illustrate the connection between the quality of the water source and cholera cases. His data convinced the local council to disable the water pump, which promptly ended the outbreak.

The Metropolis Water Act introduced the regulation of the water supply companies in London, including minimum standards of water quality for the first time. The Act "made provision for securing the supply to the Metropolis of pure and wholesome water", and required that all water be "effectually filtered" from 31 December 1855. This was followed up with legislation for the mandatory inspection of water quality, including comprehensive chemical analyses, in 1858. This legislation set a worldwide precedent for similar state public health interventions across Europe. The Metropolitan Commission of Sewers was formed at the same time, water filtration was adopted throughout the country, and new water intakes on the Thames were established above Teddington Lock. Automatic pressure filters, where the water is forced under pressure through the filtration system, were innovated in 1899 in England.

Water Chlorination

John Snow was the first to successfully use chlorine to disinfect the water supply in

Soho that had helped spread the cholera outbreak. William Soper also used chlorinated lime to treat the sewage produced by typhoid patients in 1879.

In a paper published in 1894, Moritz Traube formally proposed the addition of chloride of lime (calcium hypochlorite) to water to render it “germ-free.” Two other investigators confirmed Traube’s findings and published their papers in 1895. Early attempts at implementing water chlorination at a water treatment plant were made in 1893 in Hamburg, Germany and in 1897 the city of Maidstone England was the first to have its entire water supply treated with chlorine.

Permanent water chlorination began in 1905, when a faulty slow sand filter and a contaminated water supply led to a serious typhoid fever epidemic in Lincoln, England. Dr. Alexander Cruickshank Houston used chlorination of the water to stem the epidemic. His installation fed a concentrated solution of chloride of lime to the water being treated. The chlorination of the water supply helped stop the epidemic and as a precaution, the chlorination was continued until 1911 when a new water supply was instituted.

The first continuous use of chlorine in the United States for disinfection took place in 1908 at Boonton Reservoir (on the Rockaway River), which served as the supply for Jersey City, New Jersey. Chlorination was achieved by controlled additions of dilute solutions of chloride of lime (calcium hypochlorite) at doses of 0.2 to 0.35 ppm. The treatment process was conceived by Dr. John L. Leal and the chlorination plant was designed by George Warren Fuller. Over the next few years, chlorine disinfection using chloride of lime were rapidly installed in drinking water systems around the world.

The technique of purification of drinking water by use of compressed liquefied chlorine gas was developed by a British officer in the Indian Medical Service, Vincent B. Nesfield, in 1903. According to his own account:

It occurred to me that chlorine gas might be found satisfactory ... if suitable means could be found for using it The next important question was how to render the gas portable. This might be accomplished in two ways: By liquefying it, and storing it in lead-lined iron vessels, having a jet with a very fine capillary canal, and fitted with a tap or a screw cap. The tap is turned on, and the cylinder placed in the amount of water required. The chlorine bubbles out, and in ten to fifteen minutes the water is absolutely safe. This method would be of use on a large scale, as for service water carts.

U.S. Army Major Carl Rogers Darnall, Professor of Chemistry at the Army Medical School, gave the first practical demonstration of this in 1910. Shortly thereafter, Major William J. L. Lyster of the Army Medical Department used a solution of calcium hypochlorite in a linen bag to treat water. For many decades, Lyster’s method remained the standard for U.S. ground forces in the field and in camps, implemented in the form of the familiar Lyster Bag (also spelled Lister Bag). This work became the basis for present day systems of municipal water *purification*.

Portable Water Purification

Portable water purification devices – better described as point-of-use (POU) water treatment systems and field water disinfection techniques – are self-contained, usually hand-carried units used by recreational enthusiasts, military personnel, survivalists and people without access to water supply services in developing countries for water purification when they need to obtain drinking water from untreated sources (e.g. rivers, lakes, groundwater etc.). These personal devices and methods aim to render water potable (i.e. safe and palatable for drinking purposes - without disease-causing pathogens).

Techniques include heat (including boiling), filtration, activated charcoal absorption, chemical disinfection (e.g. chlorination, iodine, ozonation, etc.), ultraviolet purification (including SODIS), distillation (including solar distillation), and flocculation. Often these are used in combination. Many commercial portable water purification systems or chemical additives are available for hiking, camping, and other travel in remote areas.

Portable water purification methods are also commonly used at household level in developing countries when the source of drinking water is not suitable for drinking (also called "unimproved water source") - an approach called self-supply of water.

Drinking Water Hazards

Untreated water may contain potentially pathogenic agents, including protozoa, bacteria, viruses, and some larvae of higher-order parasites such as liver flukes and roundworms. Chemical pollutants such as pesticides, heavy metals and synthetic organics may be present. Other components may affect taste, odour and general aesthetic qualities, including turbidity from soil or clay, colour from humic acid or microscopic algae, odours from certain type of bacteria, particularly Actinomycetes which produce geosmin, and saltiness from brackish or sea water.

Common metallic contaminants such as copper and lead can be treated by increasing the pH using soda ash or lime, which precipitates such metals. Careful decanting of the clear water after settlement or the use of filtration provides acceptably low levels of metals. Water contaminated by aluminium or zinc cannot be treated in this way using a strong alkali as higher pHs re-dissolve the metal salts. Salt is difficult to remove except by reverse osmosis or distillation.

Most portable treatment processes focus on mitigating human pathogens for safety and removing particulates matter, tastes and odours. Significant pathogens commonly present in the developed world include Giardia, Cryptosporidium, Shigella, hepatitis A virus, Escherichia coli, and enterovirus. In less developed countries there may be risks from Cholera and Dysentery organisms and a range of tropical enteroparasites.

Giardia lamblia and *Cryptosporidium spp.*, both of which cause diarrhea are common pathogens. In backcountry areas of the United States and Canada they are sometimes present in sufficient quantity that water treatment is justified for backpackers, although this has created some controversy. In Hawaii and other tropical areas, *Leptospira spp.* are another possible problem.

Turbid creek water caused by heavy rains.

Less commonly seen in developed countries are organisms such as *Vibrio cholerae* which causes cholera and various strains of *Salmonella* which cause typhoid and para-typhoid diseases. Pathogenic viruses may also be found in water. The larvae of flukes are particularly dangerous in area frequented by sheep, deer, or cattle. If such microscopic larvae are ingested, they can form potentially life-threatening cysts in the brain or liver. This risk extends to plants grown in or near water including the commonly eaten watercress.

In general, more human activity up stream (i.e. the larger the stream/river) the greater the potential for contamination from sewage effluent, surface runoff, or industrial pollutants. Groundwater pollution may occur from human activity (e.g. on-site sanitation systems or mining) or might be naturally occurring (e.g. from arsenic in some regions of India and Bangladesh). Water collected as far upstream as possible above all known or anticipated risks of pollution poses the lowest risk of contamination and is best suited to portable treatment methods.

Techniques

31st Marine Expeditionary Unit (MEU) Service Support Group 31

Not all techniques by themselves will mitigate all hazards. Although flocculation followed by filtration has been suggested as best practice this is rarely practicable without the ability to carefully control pH and settling conditions. Ill-advised used of alum as a flocculant can lead to unacceptable levels of aluminium in the water so treated. If water is to be stored, halogens offer extended protection.

Heat (Boiling)

Heat kills disease-causing micro-organisms, with higher temperatures and/or duration required for some pathogens. Sterilization of water (killing all living containments) is not necessary to make water safe to drink; one only needs to render harmless enteric (intestinal) pathogens. Boiling does not remove most pollutants and has the potential to concentrate toxins if continued over an extended period. Boiling does not leave any residual protection.

The WHO states bringing water to rolling boil then naturally cooling is sufficient to inactivate pathogenic bacteria, viruses and protozoa.

The CDC recommends a rolling boil for 1 minute. At high elevations, though, the boiling point of water drops. At altitudes greater than 6,562 feet (2000 meters) boiling should continue for 3 minutes.

All bacterial pathogens are quickly killed above 60 °C (140 °F), therefore, although boiling is not necessary to make the water safe to drink, the time taken to heat the water to boiling is usually sufficient to reduce bacterial concentrations to safe levels. Encysted protozoan pathogens may require higher temperatures to remove any risk.

Boiling is not always necessary nor sometimes enough. Pasteurization where *enough* pathogens are killed typically occurs at 63 °C for 30 minutes or 72 °C for 15 seconds. Certain pathogens must be heated above boiling (e.g. botulism – *Clostridium botulinum* requires 118 °C (244 °F), most endospores require 120 °C (248 °F), and prions even higher). Higher temperatures may be achieved with a pressure cooker. Heat combined with ultraviolet light (UV), such as SODIS method, reduces the necessary temperature + duration.

Filtration

Portable pump filters are commercially available with ceramic filters that filter 5,000 to 50,000 litres per cartridge, removing pathogens down to the 0.2–0.3 micrometer (μm) range. Some also utilize activated charcoal filtering. Most filters of this kind remove most bacteria and protozoa, such as *Cryptosporidium* and *Giardia lamblia,* but not viruses except for the very largest of 0.3 μm and larger diameters, so disinfection by chemicals or ultraviolet light is still required after filtration. It is worth noting that not all bacteria are removed by 0.2 μm pump filters; for example, strands of thread-like *Leptospira spp.* (which can cause leptospirosis) are thin enough to pass through a 0.2 μm filter. Effective chemical additives to address shortcomings in pump filters include chlorine, chlorine dioxide, iodine, and sodium hypochlorite (bleach). There have been polymer and ceramic filters on the market that incorporated iodine post-treatment in their filter elements to kill viruses and the smaller bacteria that cannot be filtered out, but most have disappeared due to the unpleasant taste imparted to the water, as well as possible adverse health effects when iodine is ingested over protracted periods.

While the filtration elements may do an excellent job of removing most bacteria and fungi contaminants from drinking water when new, the elements themselves can become colonization sites. In recent years some filters have been enhanced by bonding silver metal nanoparticles to the ceramic element and/or to the activated charcoal to suppress growth of pathogens.

Small, hand-pumped reverse osmosis filters were originally developed for the military in the late 1980s for use as survival equipment, for example, to be included with inflatable rafts on aircraft. Civilian versions are available. Instead of using the static pressure of a water supply line to force the water through the filter, pressure is provided by a hand-operated pump, similar in function and appearance to a mechanic's grease gun. These devices can generate drinkable water from seawater.

The Portable Aqua Unit for Lifesaving (short PAUL) is a portable ultrafiltration-based membrane water filter for humanitarian aid. It allows the decentralized supply of clean water in emergency and disaster situations for about 400 persons per unit per day. The filter is designed to function with neither chemicals nor energy nor trained personnel.

Activated Charcoal Adsorption

Portable water purification unit used by International Red Cross and Red Crescent.

Granular activated carbon filtering utilizes a form of activated carbon with a high surface area, and adsorbs many compounds, including many toxic compounds. Water passing through activated carbon is commonly used in concert with hand pumped filters to address organic contamination, taste, or objectionable odors. Activated carbon filters aren't usually used as the primary purification techniques of portable water purification devices, but rather as secondary means to complement another purification technique. It is most commonly implemented for pre- or post-filtering, in a separate step than ceramic filtering, in either case being implemented prior to the addition of chemical disinfectants used to control bacteria or viruses that filters cannot remove. Activated charcoal can remove chlorine from treated water, removing any residual protection remaining in the water protecting against pathogens, and should not, in general, be used without careful thought after chemical disinfection treatments in portable water purification processing. Ceramic/Carbon Core filters with a 0.5 μm or smaller pore size are excellent for removing bacteria and cysts while also removing chemicals.

Chemical Disinfection with Halogens

Chemical disinfection with halogens, chiefly chlorine and iodine, results from oxidation of essential cellular structures and enzymes. The primary factors that determine the rate and proportion of microorganisms killed are the residual or available halogen concentration and the exposure time. Secondary factors are pathogen species, water temperature, pH, and organic contaminants. In field-water disinfection, use of concentrations of 1–16 mg/L for 10–60 min is generally effective. Of note, Cryptosporidium oocysts, likely Cyclospora species, Ascaris eggs are extremely resistant to halogens and field inactivation may not be practical with bleach and iodine.

Iodine

Iodine used for water purification is commonly added to water as a solution, in crystallized form, or in tablets containing tetraglycine hydroperiodide that release 8 mg of iodine per tablet adaptation to chronic tetraglycine hydroperiodide. The iodine kills many, but not all, of the most common pathogens present in natural fresh water sources. Carrying iodine for water purification is an imperfect but lightweight solution for those in need of field purification of drinking water. Kits are available in camping stores that include an iodine pill and a second pill (vitamin C or ascorbic acid) that will remove the iodine taste from the water after it has been disinfected. The addition of vitamin C, in the form of a pill or in flavored drink powders, precipitates much of the iodine out of the solution, so it should not be added until the iodine has had sufficient time to work. This time is 30 minutes in relatively clear, warm water, but is considerably longer if the water is turbid or cold. Iodine treated drinking water, treated with tablets containing tetraglycine hydroperiodide, also reduces the uptake of radioactive iodine in human subjects to only 2% of the value it would otherwise be. This could be an important factor worthy of consideration for treating water in a recent post nuclear event survival situation, where radioactive iodine ingestion is a concern for internal radiotoxicity. If the iodine has precipitated out of the solution, then the drinking water has less available iodine in the solution. Also the amount of iodine in one tablet is not sufficient to block uptake. Tetraglycine hydroperiodide maintains its effectiveness indefinitely before the container is opened; although some manufacturers suggest not using the tablets more than three months after the container has initially been opened, the shelf life is in fact very long provided that the container is resealed immediately after each time it is opened.

Iodine should be allowed at least 30 minutes to kill Giardia.

Iodine Crystals

A potentially lower cost alternative to using iodine-based water purification tablets is the use of iodine crystals. A small amount of water is poured into a small glass bottle (with a capacity of approximately 1 ounce or 30 cubic centimeters), containing approximately 0.25 ounces (7 grams) of iodine crystals, and then shaken vigorously to produce a saturated solution of iodine. At 25 °C (77 °F), 12.5 cubic centimeters of this solution is added to one liter of water to be disinfected. In 15 minutes the water may be consumed. At a temperature of 20 °C (68 °F) let the water to be consumed stand for 20 minutes. At a temperature of 3 °C (37 °F) let the water stand for 30 minutes or use 25 cubic centimeters of saturated solution per liter. (Do not store the saturated iodine solution in a plastic bottle as the iodine vapor will pass through the plastic and corrode steel that is in close proximity.) Note also that this method may not be adequate in killing Giardia cysts in cold water. One solution is to warm the water to be consumed to 20 °C (68 °F) before treatment. Take care not to consume iodine crystals; the amount of iodine in the saturated solution may be two to four times the lethal dose. Care must be taken to prevent the small glass bottle of iodine crystals covered with water from freezing in cold climates. An advantage of using iodine crystals is that only a small amount of iodine is dissolved from the iodine crystals at each use, giving this method of treating water a capability for treating very large amounts of water, around 2,000 liters (500 gallons), with but a small bottle of crystals. Unlike tetraglycine hydroperiodide tablets, iodine crystals have essentially an unlimited shelf life as long as they are not exposed to air for long periods of time and are kept under water. (Iodine crystals will sublimate if exposed to air for long periods of time.) The large quantity of water that can be purified with iodine crystals at low cost makes this technique especially cost effective for point of use or emergency water purification methods intended for use longer than the shelf life of tetraglycine hydroperiodide.

Halazone Tablets

Chlorine-based halazone tablets were formerly popularly used for portable water purification. Chlorine in water is more than three times more effective as a disinfectant against *Escherichia coli* than iodine. Halazone tablets were thus commonly used during World War II by U.S. soldiers for portable water purification, even being included in accessory packs for C-rations until 1945.

The primary limitation of halazone tablets was the very short usable life of opened bottles, typically 3 days or less, unlike iodine-based tablets which have a usable open bottle life of 3 months. Sodium dichloroisocyanurate (NaDCC) has largely displaced halazone tablets for the few remaining chlorine-based water purification tablets available today. It is compressed with effervescent salts, usually adipic acid and sodium bicarbonate, to form rapidly dissolving tablets, diluted to 10 parts per million available chlorine (ppm av. cl) when drinking water is mildly contaminated and 20ppm when visibly contaminated.

Chlorine bleach tablets give a more stable platform for disinfecting the water than liquid bleach (sodium hypochlorite) as the liquid version tends to degrade with age and give unregulated results unless assays are carried out – not practical on the spot. Still, despite chlorine-based halazone tablets falling from favor for portable water purification, chlorine-based bleach may nonetheless safely be used for short-term emergency water disinfection. Two drops of unscented 5% bleach can be added per liter or quart of clear water, then allowed to stand covered for 30 to 60 minutes. After this treatment, the water may be left open to reduce the chlorine smell and taste. Guidelines are available online for effective emergency use of bleach to render unsafe water potable.

The Centers for Disease Control & Prevention (CDC) and Population Services International (PSI) promote a similar product (a 0.5% - 1.5% sodium hypochlorite solution) as part of their Safe Water System (SWS) strategy. The product is sold in developing countries under local brand names specifically for the purpose of disinfecting drinking water.

Bleach

Common bleach including calcium hypochlorite ($Ca[OCl]_2$) and sodium hypochlorite ($NaOCl$) are common, well-researched, low-cost oxiders.

The EPA recommends 2 drops of 8.25% sodium hypochlorite solution (regular, unscented chlorine bleach) mixed per 1 quart/liter water and let stand 30 minutes. 2 drops of 5% solution also suffices. Double the amount of bleach if the water is cloudy, colored, or very cold. Afterwards, the water should have a slight chlorine odor, if not repeat the dosage and let stand for another 15 minutes before use.

Neither chlorine (e.g., bleach) nor iodine alone is considered completely effective against *Cryptosporidium*, although they are partially effective against *Giardia*. Chlorine is considered slightly better against the latter. A more complete field solution that includes chemical disinfectants is to first filter the water, using a 0.2 μm ceramic cartridge pumped filter, followed by treatment with iodine or chlorine, thereby filtering out cryptosporidium, Giardia, and most bacteria, along with the larger viruses, while also using chemical disinfectant to address smaller viruses and bacteria that the filter cannot remove. This combination is also potentially more effective in some cases than even using portable electronic disinfection based on UV treatment.

Chlorine Dioxide

Chlorine dioxide can come from tablets or be created by mixing two chemicals together (e.g. Aqau Mira drops). Chlorine dioxide also leaves an unpleasant taste in the water, but nowhere near as bad as iodine. The cost of chlorine dioxide treatment is about four times higher than the cost of iodine treatment.

Mixed Oxidant (MiOx)

A simple brine {salt + water} solution in an electrolytic reaction produces a powerful mixed oxidant disinfectant (mostly chlorine in the form of hypochlorous acid (HOCl) and some peroxide, ozone, chlorine dioxide) to inactivate viruses, bacteria, Giardia and Cryptosporidium killing 99.9% of all organisms without the need to filter. MiOx is considered stronger than chlorine.

Chlorine Tablets (NaDCC)

Sodium dichloroisocyanurate or Troclosene Sodium more commonly shortened as NaDCC, is a form of chlorine used for disinfection. It is used by all major NGO's such as UNICEF to treat water in emergencies, and widely by social marketing organisations for household water treatment where household sources of water may not be safe.

NaDCC tablets are available in a range of concentrations to treat differing volumes of water to give the World Heath Organisation's recommended 5ppm available chlorine. They are effervescent tablets allowing the tablet to dissolve in a matter of minutes.

Other Chemical Disinfection Additives

Silver Ion Tablets

An alternative to iodine-based preparations in some usage scenarios are silver ion/chlorine dioxide-based tablets or droplets. These solutions may disinfect water more effectively than iodine-based techniques while leaving hardly any noticeable taste in the water in some usage scenarios. Silver ion/chlorine dioxide-based disinfecting agents will kill *Cryptosporidium* and *Giardia*, if utilized correctly. The primary disadvantage of silver ion/chlorine dioxide-based techniques is the long purification times (generally 30 minutes to 4 hours, depending on the formulation used). Another concern is the possible deposition and accumulation of silver compounds in various body tissues leading to a rare condition called argyria that results in a permanent, disfiguring, bluish-gray pigmentation of the skin, eyes, and mucous membranes.

Hydrogen Peroxide

One recent study has found that the wild Salmonella which would reproduce quickly during subsequent dark storage of solar-disinfected water could be controlled by the addition of just 10 parts per million of hydrogen peroxide.

Ultraviolet Purification

Ultraviolet (UV) light induces the formation of covalent linkages on DNA and thereby prevents microbes from reproducing. Without reproduction, the microbes become far less dangerous. Germicidal UV-C light in the short wavelength range of 100–280 nm acts on thymine, one of the four base nucleotides in DNA. When a germicidal UV photon is absorbed by a thymine molecule that is adjacent to another thymine within the DNA strand, a covalent bond or dimer between the molecules is created. This thymine dimer prevents enzymes from "reading" the DNA and copying it, thus neutering the microbe. Still, there are limits to this technology. Water turbidity (i.e., the amount of suspended & colloidal solids contained in the water to be treated) must be low, such that the water is clear, for UV purification to work well - thus a pre-filter step might be necessary.

A concern with UV portable water purification is that some pathogens are hundreds of times less sensitive to UV light than others. Protozoan cysts were once believed to be among the least sensitive, however recent studies have proved otherwise, demonstrating that both Cryptosporidium and Giardia are deactivated by a UV dose of just 6 mJ/cm^2 However, EPA regulations and other studies show that it is viruses that are the limiting factor of UV treatment, requiring a 10-30 times greater dose of UV light than *Giardia* or *Cryptosporidium*. Studies have shown that UV doses at the levels provided by common portable UV units are effective at killing *Giardia* and that there was no evidence of repair and reactivation of the cysts.

Water treated with UV still has the microbes present in the water, only with their means for reproduction turned "off". In the event that such UV-treated water containing neutered microbes is exposed to visible light (specifically, wavelengths of light over 330-500 nm) for any significant period of time, a process known as photo reactivation can take place, where the possibility for repairing the damage in the bacteria's reproduction DNA arises, potentially rendering them once more capable of reproducing and causing disease. UV-treated water must therefore not be exposed to visible light for any significant period of time after UV treatment, before consumption, to avoid ingesting reactivated and dangerous microbes.

Recent developments in semiconductor technology allows for the development of UV-C Light Emitting Diodes (LEDs). UV-C LED systems address disadvantages of mercury-based technology, namely: power-cycling penalties, high power needs, fragility, warm-up time, and mercury content.

Ozone Water Disinfection

In ozone water disinfection, microbes are destroyed by ozone gas (O_3) provided by an ozone generator. Common in Europe, ozone gas is now becoming widely adopted in the United States. It is emerging across a wide array of industries; from municipal water treatment plants, to food processing plants, to healthcare organizations. It is being adopted due to its ability to sanitize water and surfaces without wasting water, and because there are no by-products. When its job is done, ozone gas quickly degrades into oxygen. Ozone is more effective than chlorine in destroying viruses and bacteria.

In 1990, the Organic Foods Production Act (OFPA) identified aqueous ozone as a substance that is allowed for use in organic crop and livestock production. In 1997, it was approved by the FDA as an antimicrobial agent for use on food. In 2002, the FDA approved ozone for use on food contact areas and directly on food with its Generally Regarded as Safe ("GRAS") designation.

Ozone is most commonly created by a process called "corona discharge", which causes oxygen molecules (O_2) to temporarily re-combine into ozone (O_3). This gas is very unstable, and the 3rd oxygen molecule reacts with pathogens by penetrating the cell walls of bacteria and viruses. This destroys the organisms.

Ozone is effective against pollutants for the same reason; it will react with long-chain carbon (organic) molecules, and break them down into less complex (and typically less harmful) molecules through oxidation.

Advances in ozone generation techniques, coupled with filtration, make this a viable new portable water purification method.

Solar Water Disinfection

In solar water disinfection (SODIS), microbes are destroyed by temperature and UVA radiation provided by the sun. Water is placed in a transparent plastic PET bottle or plastic bag, oxygenated by shaking partially filled capped bottles prior to filling the bottles all the way, and left in the sun for 6–24 hours atop a reflective surface.

Solar Distillation

Solar distillation relies on sunlight to warm and evaporate the water to be purified which then condenses and trickles into a container. In theory, a solar (condensation) still removes all pathogens, salts, metals, and most chemicals but in field practice the lack of clean components, easy contact with dirt, improvised construction, and disturbances result in cleaner, yet contaminated water.

Homemade Water Filters

Water filters can be made on-site using local materials such as sand and charcoal (e.g. from firewood burned in a special way). These filters are sometimes used by soldiers and outdoor enthusiasts. Due to their low cost they can be made and used by anyone. The reliability of such systems is highly variable. Such filters can do little, if anything, to mitigate germs and other harmful constituents and can give a false sense of security that the water so produced is potable. Water processed through an improvised filter should undergo secondary processing such as boiling to render it safe for consumption.

Prevention of Water Contamination

Human disease/illness usually come from other humans, thus human-derived materials (feces, medical waste, wash water, lawn chemicals, gasoline engines, garbage, etc.) should be kept far away from water sources. For example, bury human waste well away (>60 meters/200 feet) from water sources to reduce contamination. In some wilderness areas it is recommended that all waste be packed up and carted out to a properly designated disposal point.

Desalination

Desalination is a process that extracts minerals from saline water. More generally, desalination refers to the removal of salts and minerals from a target substance, as in soil desalination, which is an issue for agriculture.

Saltwater is desalinated to produce water suitable for human consumption or irrigation. One by-product of desalination is salt. Desalination is used on many seagoing ships and submarines. Most of the modern interest in desalination is focused on cost-effective provision of fresh water for human use. Along with recycled wastewater, it is one of the few rainfall-independent water sources.

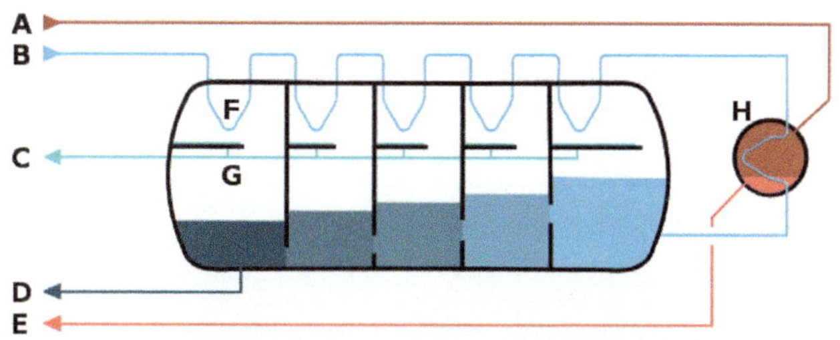

Schematic of a multistage flash desalinator
A – steam in B – seawater in C – potable water out D – waste out
E – steam out F – heat exchange G – condensation collection H – brine heater

Due to its energy consumption, desalinating sea water is generally more costly than fresh water from rivers or groundwater, water recycling and water conservation. However, these alternatives are not always available and depletion of reserves is a critical problem worldwide. Currently, approximately 1% of the world's population is dependent on desalinated water to meet daily needs, but the UN expects that 14% of the world's population will encounter water scarcity by 2025.

Desalination is particularly relevant in dry countries such as Australia, which traditionally have relied on collecting rainfall behind dams for water.

According to the International Desalination Association, in June 2015, 18,426 desalination plants operated worldwide, producing 86.8 million cubic meters per day, providing water for 300 million people. This number increased from 78.4 million cubic meters in 2013, a 10.71% increase in 2 years. The single largest desalination project is Ras Al-Khair in Saudi Arabia, which produced 1,025,000 cubic meters per day in 2014, although this plant is expected to be surpassed by a plant in California. Kuwait produces a higher proportion of its water than any other country, totaling 100% of its water use.

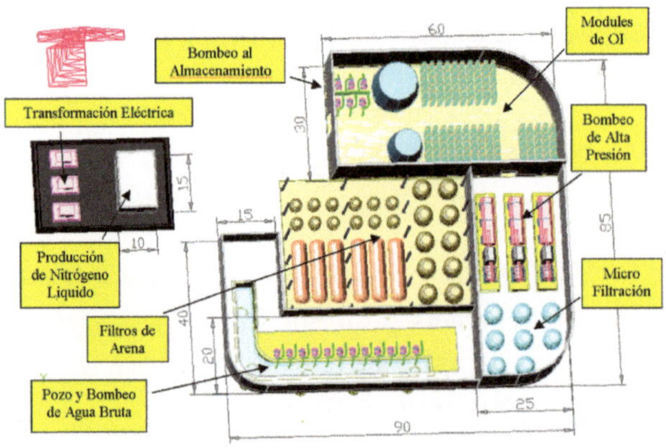

Plan of a typical reverse osmosis desalination plant.

Methods

Reverse osmosis desalination plant in Barcelona, Spain.

There are several methods. Each has advantages and disadvantages.

Vacuum Distillation

The traditional process used in these operations is vacuum distillation—essentially boiling it to leave impurities behind. In desalination, atmospheric pressure is reduced, thus lowering the required temperature needed. Liquids boil when the vapor pressure equals the ambient pressure and vapor pressure increases with temperature. Thus, because of the reduced temperature, low-temperature "waste" heat from electrical power generation or industrial processes can be employed.

Multi-Stage Flash Distillation

Water is evaporated and separated from sea water through multi-stage flash distillation, which is a series of flash evaporations. Each subsequent flash process utilizes energy released from the condensation of the water vapor from the previous step.

Multiple Effect Distillation

Multiple effect distillation (MED) works through a series of steps called "effects". Incoming water is sprayed onto vertically or, more commonly, horizontally oriented pipes which are then heated to generate steam. The steam is then used to heat the next batch of incoming sea water. To increase efficiency, the steam used to heat the sea water can be taken from nearby power plants. Although this method is the most thermodynamically efficient, a few limitations exist such as a max temperature and max number of effects.

Vapor Compression Distillation

Vapor compression distillation involves using either a mechanical compressor or a jet stream to compress the vapor present above the liquid. The compressed vapor is then used to provide the heat needed for the evaporation of the rest of the sea water. Since this system only requires power, it is more efficient if kept at a small scale.

Reverse Osmosis

The principal competing process uses membranes to desalt saline water, principally applying reverse osmosis (RO). The RO membrane processes use semipermeable membranes and applied pressure (on the membrane feed side) to preferentially induce water permeation through the membrane while rejecting salts. Reverse osmosis plant membrane systems typically use less energy than thermal desalination processes. Desalination processes are driven by either thermal (e.g., distillation) or electrical (e.g., RO) as the primary energy types. Energy cost in desalination processes varies considerably depending on water salinity, plant size and process type. At present the cost of seawater desalination, for example, is higher than traditional water sources, but it is expected that costs will continue to decrease with technology improvements that include, but are not limited to, reduction in plants footprint, improvements to plant operation and optimization, more effective feed pretreatment, and lower cost energy sources.

Freezing

Freezing involves partially freezing the seawater and removing the ice to then clean and melt back down into now clean water. In the freezing process, water is separated from the salt in order for the water to crystallize properly.

Solar Evaporation

Solar evaporation mimics the natural water cycle, in which the sun heats the sea water enough for evaporation to occur. After evaporation, the water vapor is condensed onto a cool surface.

Electrodialysis/Electrodialysis Reversal

Electrodialysis utilizes electric potential to move the salts through a membrane.

Considerations and criticism

Energy consumption

Energy consumption of seawater desalination has reached as low as 3 kWh/m^3, including pre-filtering and ancillaries, similar to the energy consumption of other fresh water supplies transported over large distances, but much higher than local fresh water supplies that use 0.2 kWh/m^3 or less.

A minimum energy consumption for seawater desalination of around 1 kWh/m^3 has been determined, excluding prefiltering and intake/outfall pumping. Under 2 kWh/m^3 has been achieved with reverse osmosis membrane technology, leaving limited scope for further energy reductions.

Supplying all US domestic water by desalination would increase domestic energy consumption by around 10%, about the amount of energy used by domestic refrigerators. Domestic consumption is a relatively small fraction of the total water usage.

Energy consumption of seawater desalination methods.

Desalination Method >>	**Multi-stage Flash MSF**	**Multi-Effect Distillation MED**	**Mechanical Vapor Compression MVC**	**Reverse Osmosis RO**
Electrical energy (kWh/m^3)	4–6	1.5–2.5	7–12	3–5.5
Thermal energy (kWh/m^3)	50–110	60–110	None	None
Electrical equivalent of thermal energy (kWh/m^3)	9.5–19.5	5–8.5	None	None
Total equivalent electrical energy (kWh/m^3)	13.5–25.5	6.5–11	7–12	3–5.5

"Electrical equivalent" refers to the amount of electrical energy that could be generated using a given quantity of thermal energy and appropriate turbine generator. These calculations do not include the energy required to construct or refurbish items consumed in the process.

Cogeneration

Cogeneration is generating excess heat and electricity generation from a single process. Cogeneration can provide usable heat for desalination in an integrated, or "dual-purpose", facility where a power plant provides the energy for desalination. Alternatively, the facility's energy production may be dedicated to the production of potable water (a stand-alone facility), or excess energy may be produced and incorporated into the energy grid. Cogeneration takes various forms, and theoretically any form of energy production could be used. However, the majority of current and planned cogeneration desalination plants use either fossil fuels or nuclear power as their source of energy. Most plants are located in the Middle East or North Africa, which use their petroleum resources to offset limited water resources. The advantage of dual-purpose facilities is they can be more efficient in energy consumption, thus making desalination more viable.

The Shevchenko BN350, a nuclear-heated desalination unit.

The current trend in dual-purpose facilities is hybrid configurations, in which the permeate from reverse osmosis desalination is mixed with distillate from thermal desalination. Basically, two or more desalination processes are combined along with power production. Such facilities have been implemented in Saudi Arabia at Jeddah and Yanbu.

A typical Supercarrier in the US military uses nuclear power to desalinate 400,000 US gallons (1,500,000 l; 330,000 imp gal) of water per day.

Economics

Costs of desalinating sea water (infrastructure, energy, and maintenance) are generally higher than fresh water from rivers or groundwater, water recycling, and water conservation, but alternatives are not always available. Desalination costs in 2013 ranged from US$0.45 to $1.00/cubic metre ($US2 to 4/kgal). (1 cubic meter is about 264 gallons.) More than half of the cost comes directly from energy cost, and since energy prices are very volatile, actual costs can vary substantially.

The cost of untreated fresh water in the developing world can reach US$5/cubic metre.

Average water consumption and cost of supply by sea water desalination at US$1 per cubic metre(±50%)			
Area	**Consumption USgal/person/ day**	**Consump-tion litre/ person/day**	**Desalinated Water Cost US$/person/day**
USA	100	378	0.38
Europe	50	189	0.19
Africa	15	57	0.06
UN recommended minimum	13	49	0.05

Factors that determine the costs for desalination include capacity and type of facility, location, feed water, labor, energy, financing and concentrate disposal. Desalination stills control pressure, temperature and brine concentrations to optimize efficiency. Nuclear-powered desalination might be economical on a large scale.

While noting costs are falling, and generally positive about the technology for affluent areas in proximity to oceans, a 2004 study argued, "Desalinated water may be a solution for some water-stress regions, but not for places that are poor, deep in the interior of a continent, or at high elevation. Unfortunately, that includes some of the places with biggest water problems.", and, "Indeed, one needs to lift the water by 2,000 m (6,600 ft), or transport it over more than 1,600 km (990 mi) to get transport costs equal to the desalination costs. Thus, it may be more economical to transport fresh water from somewhere else than to desalinate it. In places far from the sea, like New Delhi, or in high places, like Mexico City, transport costs could match desalination costs. Desalinated water is also expensive in places that are both somewhat far from the sea and somewhat high, such as Riyadh and Harare. By contrast in other locations transport costs are much less, such as Beijing, Bangkok, Zaragoza, Phoenix, and, of course, coastal cities like Tripoli." After desalination at Jubail, Saudi Arabia, water is pumped 200 mi (320 km) inland to Riyadh. For coastal cities, desalination is increasingly viewed as a competitive choice.

In 2014, the Israeli facilities of Hadera, Palmahim, Ashkelon, and Sorek were desalinizing water for less than US$0.40 per cubic meter. As of 2006, Singapore was desalinating water for US$0.49 per cubic meter. The city of Perth began operating a reverse osmosis seawater desalination plant in 2006. A desalination plant now operates in Sydney, and the Wonthaggi desalination plant was under construction in Wonthaggi, Victoria.

The Perth desalination plant is powered partially by renewable energy from the Emu Downs Wind Farm. A wind farm at Bungendore in New South Wales was purpose-built to generate enough renewable energy to offset the Sydney plant's energy use, mitigating concerns about harmful greenhouse gas emissions.

In December 2007, the South Australian government announced it would build a seawater desalination plant for the city of Adelaide, Australia, located at Port Stanvac. The desalination plant was to be funded by raising water rates to achieve full cost recovery.

A January 17, 2008, article in the *Wall Street Journal* stated, "In November, Connecticut-based Poseidon Resources Corp. won a key regulatory approval to build the $300 million water-desalination plant in Carlsbad, north of San Diego. The facility would produce 50,000,000 US gallons (190,000,000 l; 42,000,000 imp gal) of drinking water per day, enough to supply about 100,000 homes. As of June 2012, the cost for the desalinated water had risen to $2,329 per acre-foot. Each $1,000 per acre-foot works out to $3.06 for 1,000 gallons, or $.81 per cubic meter.

Poseidon Resources made an unsuccessful attempt to construct a desalination plant in Tampa Bay, FL, in 2001. The board of directors of Tampa Bay Water was forced to buy the plant from Poseidon in 2001 to prevent a third failure of the project. Tampa Bay Water faced five years of engineering problems and operation at 20% capacity to protect marine life. The facility reached capacity only in 2007.

In 2008, a Energy Recovery Inc. was desalinating water for $0.46 per cubic meter.

Environmental

Factors that determine the costs for desalination include capacity and type of facility, location, feed water, labor, energy, financing and concentrate disposal.

Intake

In the United States, cooling water intake structures are regulated by the Environmental Protection Agency (EPA). These structures can have the same impacts to the environment as desalination facility intakes. According to EPA, water intake structures cause adverse environmental impact by sucking fish and shellfish or their eggs into an industrial system. There, the organisms may be killed or injured by heat, physical stress, or chemicals. Larger organisms may be killed or injured when they become trapped against screens at the front of an intake structure. Alternative intake types that mitigate these impacts include beach wells, but they require more energy and higher costs.

The Kwinana Desalination Plant opened in Perth in 2007. Water there and at Queensland's Gold Coast Desalination Plant and Sydney's Kurnell Desalination Plant is withdrawn at 0.1 m/s (0.33 ft/s), which is slow enough to let fish escape. The plant provides nearly 140,000 m^3 (4,900,000 cu ft) of clean water per day.

Outflow

Desalination processes produce large quantities of brine, possibly at above ambient temperature, and contain residues of pretreatment and cleaning chemicals, their reaction byproducts and heavy metals due to corrosion. Chemical pretreatment and cleaning are a necessity in most desalination plants, which typically includes prevention of biofouling, scaling, foaming and corrosion in thermal plants, and of biofouling, suspended solids and scale deposits in membrane plants.

To limit the environmental impact of returning the brine to the ocean, it can be diluted with another stream of water entering the ocean, such as the outfall of a wastewater treatment or power plant. With medium to large power plant and desalination plants, the power plant's cooling water flow is likely to be several times larger than that of the desalination plant, reducing the salinity of the combination. Another method to dilute the brine is to mix it via a diffuser in a mixing zone. For example, once a pipeline containing the brine reaches the sea floor, it can split into many branches, each releasing brine gradually through small holes along its length. Mixing can be combined with power plant or wastewater plant dilution.

Brine is denser than seawater and therefore sinks to the ocean bottom and can damage the ecosystem. Careful reintroduction can minimize this problem. Typical ocean conditions allow for rapid dilution, thereby minimizing harm.

Alternatives Without Pollution

Some methods of desalination, particularly in combination with evaporation ponds, solar stills, and condensation trap (solar desalination), do not discharge brine. They do not use chemicals or burn fossil fuels. They do not work with membranes or other critical parts, such as components that include heavy metals, thus do not produce toxic waste (and high maintenance).

A new approach that works like a solar still, but on the scale of industrial evaporation ponds is the integrated biotectural system. It can be considered "full desalination" because it converts the entire amount of saltwater intake into distilled water. One of the advantages of this system is the feasibility for inland operation. Standard advantages also include no air pollution and no temperature increase of endangered natural water bodies from power plant cooling-water discharge. Another important advantage is the production of sea salt for industrial and other uses. As of 2015, 50% of the world's sea salt production relies on fossil energy sources.

Alternatives to Desalination

Increased water conservation and efficiency remain the most cost-effective approaches in areas with a large potential to improve the efficiency of water use practices. Wastewater reclamation provides multiple benefits over desalination. Urban runoff and storm water capture also provide benefits in treating, restoring and recharging groundwater.

A proposed alternative to desalination in the American Southwest is the commercial importation of bulk water from water-rich areas either by oil tankers converted to water carriers, or pipelines. The idea is politically unpopular in Canada, where governments imposed trade barriers to bulk water exports as a result of a North American Free Trade Agreement (NAFTA) claim.

Public Health Concerns

Desalination removes iodine from water and could increase the risk of iodine deficiency disorders. Israeli researchers claimed a possible link between seawater desalination and iodine deficiency, finding deficits among euthyroid adults exposed to iodine-poor water concurrently with an increasing proportion of their area's drinking water from seawater reverse osmosis (SWRO). They later found probable iodine deficiency disorders in a population reliant on desalinated seawater. A possible link of heavy desalinated water use and national iodine deficiency was suggested by Israeli researchers. They found a high burden of iodine deficiency in the general population of Israel: 62% of school-age children and 85% of pregnant women fall below the WHO's adequacy range. They also pointed out the national reliance on iodine-depleted desalinated water, the absence of a universal salt iodization program and reports of increased use of thyroid medication in Israel as a possible reasons that the population's iodine intake low. In the year that the survey was conducted, the amount of water produced from the desalination plants constitutes about 50% of the quantity of fresh water supplied for all needs and about 80% of the water supplied for domestic and industrial needs in Israel.

Other Issues

Due to the nature of the process, there is a need to place the plants on approximately 25 acres of land on or near the shoreline. In the case a plant is built inland, pipes have to be laid into the ground to allow for easy intake and outtake. However, once the pipes are laid into the ground, they have a possibility of leaking into and contaminating nearby aquifers. Aside from environmental risks, the noise generated by certain types of desalination plants can be loud.

Experimental Techniques

Other desalination techniques include:

Waste Heat

Diesel generators commonly provide electricity in remote areas. About 40%–50% of the energy output is low-grade heat that leaves the engine via the exhaust. Connecting a membrane distillation system to the diesel engine exhaust repurposes this low-grade heat for desalination. The system actively cools the diesel generator, improving its efficiency and increasing its electricity output. This results in an energy-neutral desalination solution. An example plant was commissioned by Dutch company Aquaver in March 2014 for Gulhi, Maldives.

Low-temperature Thermal

Originally stemming from ocean thermal energy conversion research, low-temperature thermal desalination (LTTD) takes advantage of water boiling at low pressure, even at ambient temperature. The system uses pumps to create a low-pressure, low-temperature environment in which water boils at a temperature gradient of 8–10 °C (46–50 °F) between two volumes of water. Cool ocean water is supplied from depths of up to 600 m (2,000 ft). This water is pumped through coils to condense the water vapor. The resulting condensate is purified water. LTTD may take advantage of the temperature gradient available at power plants, where large quantities of warm wastewater are discharged from the plant, reducing the energy input needed to create a temperature gradient.

Experiments were conducted in the US and Japan to test the approach. In Japan, a spray-flash evaporation system was tested by Saga University. In Hawaii, the National Energy Laboratory tested an open-cycle OTEC plant with fresh water and power production using a temperature difference of 20 C° between surface water and water at a depth of around 500 m (1,600 ft). LTTD was studied by India's National Institute of Ocean Technology (NIOT) in 2004. Their first LTTD plant opened in 2005 at Kavaratti in the Lakshadweep islands. The plant's capacity is 100,000 L (22,000 imp gal; 26,000 US gal)/day, at a capital cost of INR 50 million (€922,000). The plant uses deep water at a temperature of 10 to 12 °C (50 to 54 °F). In 2007, NIOT opened an experimental, floating LTTD plant off the coast of Chennai, with a capacity of 1,000,000 L (220,000 imp gal; 260,000 US gal)/day. A smaller plant was established in 2009 at the North Chennai Thermal Power Station to prove the LTTD application where power plant cooling water is available.

Thermoionic Process

In October 2009, Saltworks Technologies announced a process that uses solar or other thermal heat to drive an ionic current that removes all sodium and chlorine ions from the water using ion-exchange membranes.

Evaporation and Condensation for Crops

The Seawater greenhouse uses natural evaporation and condensation processes inside a greenhouse powered by solar energy to grow crops in arid coastal land.

Other Approaches

Adsorption-based desalination (AD) relies on the moisture absorption properties of certain materials such as Silica Gel.

Forward Osmosis

One process was commercialized by Modern Water PLC using forward osmosis, with a number of plants reported to be in operation.

Small-scale Solar

The United States, France and the United Arab Emirates are working to develop practical solar desalination. AquaDania's WaterStillar has been installed at Dahab, Egypt, and in Playa del Carmen, Mexico. In this approach, a solar thermal collector measuring two square metres can distill from 40 to 60 litres per day from any local water source – five times more than conventional stills. It eliminates the need for plastic PET bottles or energy-consuming water transport. In Central California, a startup company WaterFX is developing a solar-powered method of desalination that can enable the use of local water, including runoff water that can be treated and used again. Salty groundwater in the region would be treated to become freshwater, and in areas near the ocean, seawater could be treated.

Passarell

The Passarell process uses reduced atmospheric pressure rather than heat to drive evaporative desalination. The pure water vapor generated by distillation is then compressed and condensed using an advanced compressor. The compression process improves distillation efficiency by creating the reduced pressure in the evaporation chamber. The compressor centrifuges the pure water vapor after it is drawn through a demister (removing residual impurities) causing it to compress against tubes in the collection chamber. The compression of the vapor increases its temperature. The heat is transferred to the input water falling in the tubes, vaporizing the water in the tubes. Water vapor condenses on the outside of the tubes as product water. By combining several physical processes, Passarell enables most of the system's energy to be recycled through its evaporation, demisting, vapor compression, condensation, and water movement processes.

Geothermal

Geothermal energy can drive desalination. In most locations, geothermal desalination beats using scarce groundwater or surface water, environmentally and economically.

Nanotechnology

Nanotube membranes of higher permeability than current generation of membranes may lead to eventual reduction in the footprint of RO desalination plants. It has also been suggested that the use of such membranes will lead to reduction in the energy needed for desalination.

Hermetic, sulphonated nano-composite membranes have shown to be capable of removing a various contaminants to the parts per billion level. s, have little or no susceptibility to high salt concentration levels.

Biomimesis

Biomimetic membranes are another approach.

Electrochemical

In 2008, Siemens Water Technologies announced technology that applied electric fields to desalinate one cubic meter of water while using only a purported 1.5 kWh of energy. If accurate, this process would consume one-half the energy of other processes. As of 2012 a demonstration plant was operating in Singapore. Researchers at the University of Texas at Austin and the University of Marburg are developing more efficient methods of electrochemically mediated seawater desalination.

Freeze-thaw

Freeze-thaw desalination uses freezing to remove fresh water from frozen seawater.

Electrokinetic Shocks

A process employing electrokinetic shocks waves can be used to accomplish membraneless desalination at ambient temperature and pressure. In this process, anions and cations in salt water are exchanged for carbonate anions and calcium cations, respectively using electrokinetic shockwaves. Calcium and carbonate ions react to form calcium carbonate, which precipitates, leaving fresh water. The theoretical energy efficiency of this method is on par with electrodialysis and reverse osmosis.

Facilities

Estimates vary widely between 15,000–20,000 desalination plants producing more than 20,000 m^3/day. Micro desalination plants operate near almost every natural gas or fracking facility found in the United States.

In Nature

Evaporation of water over the oceans in the water cycle is a natural desalination process.

The formation of sea ice produces ice with little salt, much lower than in seawater.

Seabirds distill seawater using countercurrent exchange in a gland with a rete mirabile. The gland secretes highly concentrated brine stored near the nostrils above the beak. The bird then "sneezes" the brine out. As freshwater is not usually available in their environments, some seabirds, such as pelicans, petrels, albatrosses, gulls and terns, possess this gland, which allows them to drink the salty water from their environments while they are far from land.

Mangrove leaf with salt crystals

Mangrove trees grow in seawater; they secrete salt by trapping it in parts of the root, which are then eaten by animals (usually crabs). Additional salt is removed by storing it in leaves that fall off. Some types of mangroves have glands on their leaves, which work in a similar way to the seabird desalination gland. Salt is extracted to the leaf exterior as small crystals, which then fall off the leaf.

Willow trees and reeds absorb salt and other contaminants, effectively desalinating the water. This is used in artificial constructed wetlands, for treating sewage.

History

Desalination has been known to history for millennia as both a concept, and later practice, though in a limited form. The ancient Greek philosopher Aristotle observed in his work *Meteorology* that "salt water, when it turns into vapour, becomes sweet and the vapour does not form salt water again when it condenses," and also noticed that a fine wax vessel would hold potable water after being submerged long enough in seawater, having acted as a membrane to filter the salt. There are numerous other examples of experimentation in desalination throughout Antiquity and the Middle Ages, but desalination was never feasible on a large scale until the modern era.

Before the Industrial Revolution, desalination was primarily of concern to oceangoing ships, which otherwise needed to keep on board supplies of fresh water. When Protector (1779 frigate) was sold to Denmark in the 1780s (as the ship Hussaren) the desalination plant was studied and recorded in great detail.In the newly formed United States, Thomas

Jefferson catalogued heat-based methods going back to the 1500s, and formulated practical advice that was publicized to all U.S. ships on the backs of sailing clearance permits.

Significant research into improved desalination methods occurred in the United States after World War II. The Office of Saline Water was created in the United States Department of the Interior by the Saline Water Conversion Act of 1952. It was merged into the Office of Water Resources Research in 1974.

Research also took place at state universities in California, followed by development at the Dow Chemical Company and DuPont. Many studies focus on ways to optimize desalination systems.

Salinometer

A salinometer

A salinometer is a device designed to measure the salinity, or dissolved salt content, of a solution.

Since the salinity affects both the electrical conductivity and the specific gravity of a solution, a salinometer often consist of an ec meter or hydrometer and some means of converting those readings to a salinity reading. A salinometer may be calibrated in either micromhos, a unit of electrical conductivity, (usually 0-22) or else directly calibrated for salt in 'grains per gallon' (0-0.5). A typical reading on-board ship would be 2 micromhos or 0.05 grains per gallon. A reading of twice this may trigger a warning light or alarm.

Applications

Fresh water generators (Evaporators) use salinometers on the distillate discharge in order to gauge the quality of the water. Water from the evaporator can be destined for potable water supplies, so salty water is not desirable for human consumption.

In some ships, extremely high quality distillate is required for use in water-tube boilers, where salt water would be disastrous. In these ships, a salinometer is also installed on the feed system where it would alert the engineer to any salt contamination. The salinometer may switch the evaporator's output from fresh-water to feed-water tanks automatically, depending on the water quality. The *higher* quality (lower salinity) is required for the boiler feedwater, not for drinking.

Reclaimed Water

Sequence of reclamation from left: raw sewage, plant effluent, and finally reclaimed water (after several treatment steps).

Reclaimed or recycled water (also called wastewater reuse or water reclamation) is the process of converting wastewater into water that can be reused for other purposes. Reuse may include irrigation of gardens and agricultural fields or replenishing surface water and groundwater (i.e., groundwater recharge). Reused water may also be directed toward fulfilling certain needs in residences (e.g. toilet flushing), businesses, and industry, and could even be treated to reach drinking water standards. This last option is called either "direct potable reuse" or "indirect potable" reuse, depending on the approach used. Colloquially, the term "toilet to tap" also refers to potable reuse.

Reclaiming water for reuse applications instead of using freshwater supplies can be a water-saving measure. When used water is eventually discharged back into natural water sources, it can still have benefits to ecosystems, improving streamflow, nourishing plant life and recharging aquifers, as part of the natural water cycle.

Wastewater reuse is a long-established practice used for irrigation, especially in arid countries. Reusing wastewater as part of sustainable water management allows water to remain as an alternative water source for human activities. This can reduce demand and alleviate pressures on groundwater and other natural water bodies.

Background

Irrigation water is pumped from this tank which stores effluent received from a constructed wetland in Haran-Al-Awamied, Syria.

Achieving more sustainable sanitation and wastewater management will require emphasis on actions linked to resource management, such as wastewater reuse or excreta reuse that will keep valuable resources available for productive uses. This in turn supports human wellbeing and broader sustainability.

Simply stated, reclaimed water is water that is used more than one time before it passes back into the natural water cycle. Advances in wastewater treatment technology allow communities to reuse water for many different purposes. The water is treated differently depending upon the source and use of the water and how it gets delivered.

Cycled repeatedly through the planetary hydrosphere, all water on Earth is recycled water, but the terms "recycled water" or "reclaimed water" typically mean wastewater sent from a home or business through a sewer system to a wastewater treatment plant, where it is treated to a level consistent with its intended use.

The World Health Organization has recognized the following principal driving forces for wastewater reuse:

1. increasing water scarcity and stress,
2. increasing populations and related food security issues,
3. increasing environmental pollution from improper wastewater disposal, and
4. increasing recognition of the resource value of wastewater, excreta and greywater.

Water is a limiting resource, and the pressure exerted on surface and groundwater resources should be reduced or at best maintained, rather than increased, as the human population and industrial development increase. Water recycling and reuse is thus of increasing importance, not only in arid regions but also in cities and contaminated environments.

Already, the groundwater aquifers that are used by over half of the world population are being over-drafted. Reuse will continue to increase as the world's population becomes increasingly urbanized and concentrated near coastlines, where local freshwater supplies are limited or are available only with large capital expenditure. Large quantities of freshwater can be saved by wastewater reuse and recycling, reducing environmental pollution and improving carbon footprint. Reuse can be an alternative water supply option.

Types and Applications

Most of the uses of water reclamation are non potable uses such as: washing cars, flushing toilets, cooling water for power plants, concrete mixing, artificial lakes, irrigation for golf courses and public parks, and for hydraulic fracturing. Where applicable, systems run a dual piping system to keep the recycled water separate from the potable water.

The main reclaimed water applications in the world are shown below:

Categories of use	Uses
Urban uses	Irrigation of public parks, sporting facilities, private gardens, roadsides; Street cleaning; Fire protection systems; Vehicle washing; Toilet flushing; Air conditioners; Dust control.
Agricultural uses	Food crops not commercially processed; Food crops commercially processed; Pasture for milking animals; Fodder; Fibre; Seed crops; Ornamental flowers; Orchards; Hydroponic culture; Aquaculture; Greenhouses; Viticulture.
Industrial uses	Processing water; Cooling water; Recirculating cooling towers; Washdown water; Washing aggregate; Making concrete; Soil compaction; Dust control.
Recreational uses	Golf course irrigation; Recreational impoundments with/without public access (e.g. fishing, boating, bathing); Aesthetic impoundments without public access; Snowmaking.
Environmental uses	Aquifer recharge; Wetlands; Marshes; Stream augmentation; Wildlife habitat; Silviculture.
Potable uses	Aquifer recharge for drinking water use; Augmentation of surface drinking water supplies; Treatment until drinking water quality.

De Facto Wastewater Reuse (Unplanned Potable Reuse)

De facto, unacknowledged or unplanned potable reuse refers to a situation where reuse of treated wastewater is, in fact, practiced but is not officially recognized. For example, a wastewater treatment plant from one city may be discharging effluents to a river which is used as a drinking water supply for another city downstream.

Unplanned Indirect Potable Use has existed for a long time. Large towns on the River Thames upstream of London (Oxford, Reading, Swindon, Bracknell) discharge their treated sewage ("non-potable water") into the Thames, which supplies water to London downstream. In the United States, the Mississippi River serves as both the destination of sewage treatment plant effluent and the source of potable water.

Urban Reuse

- Unrestricted: The use of reclaimed water for non-potable applications in municipal settings, where public access is not restricted.
- Restricted: The use of reclaimed water for non-potable applications in municipal settings, where public access is controlled or restricted by physical or institutional barriers, such as fencing, advisory signage, or temporal access restriction.

Agricultural Reuse

There are benefits of using recycled water for irrigation, including the lower cost compared to some other sources and consistency of supply regardless of season, climatic conditions and associated water restrictions. When reclaimed water is used for irrigation in agriculture, the nutrient (nitrogen and phosphorus) content of the treated wastewater has the benefit of acting as a fertilizer. This can make the reuse of excreta contained in sewage attractive.

The irrigation water can be used in different ways on different crops:

- Food crops to be eaten raw: crops which are intended for human consumption to be eaten raw or unprocessed.
- Processed food crops: crops which are intended for human consumption not to be eaten raw but after treatment process (i.e. cooked, industrially processed).
- Non-food crops: crops which are not intended for human consumption (e.g. pastures, forage, fiber, ornamental, seed, forest and turf crops).

In developing countries, agriculture is increasingly using untreated wastewater for irrigation - often in an unsafe manner. Cities provide lucrative markets for fresh produce, so are attractive to farmers. However, because agriculture has to compete for increasingly scarce water resources with industry and municipal users, there is often no alternative for farmers but to use water polluted with urban waste directly to water their crops.

There can be significant health hazards related to using untreated wastewater in agriculture. Wastewater from cities can contain a mixture of chemical and biological pollutants. In low-income countries, there are often high levels of pathogens from excreta. In emerging nations, where industrial development is outpacing environmental regulation, there are increasing risks from inorganic and organic chemicals. The World Health Organization, in collaboration with the Food and Agriculture Organization of the United Nations (FAO) and the United Nations Environmental Program (UNEP), has developed guidelines for safe use of wastewater in 2006. These guidelines advocate a 'multiple-barrier' approach to wastewater use, for example by encouraging farmers to adopt various risk-reducing behaviours. These include ceasing irrigation a few days before harvesting to allow pathogens to die off in the sunlight, applying water carefully so it does not contaminate leaves likely to be eaten raw, cleaning vegetables with disinfectant or allowing fecal sludge used in farming to dry before being used as a human manure.

Environmental Reuse

The use of reclaimed water to create, enhance, sustain, or augment water bodies including wetlands, aquatic habitats, or stream flow is called "environmental reuse". For example, constructed wetlands fed by wastewater provide both wastewater treatment and habitats for flora and fauna.

Industrial Reuse

The use of reclaimed water to recharge aquifers that are not used as a potable water source.

Planned Potable Reuse

Planned potable reuse is publicly acknowledged as an intentional project to recycle water for drinking water. How potable reused water is delivered determines if it is called "Indirect Potable Reuse" or "Direct Potable Reuse". Both forms of reuse commonly involve a more formal public process and public consultation program than is the case with de facto or unacknowledged reuse. In 'indirect' potable reuse applications, the reclaimed wastewater is used directly or mixed with other sources.

Direct potable reuse is also called "toilet to tap".

Some water agencies reuse highly treated effluent from municipal wastewater or resource recovery plants as a reliable, drought proof source of drinking water. By using advanced purification processes, they produce water that meets all applicable drinking water standards. System reliability and frequent monitoring and testing are imperative to them meeting stringent controls.

The water needs of a community, water sources, public health regulations, costs, and the types of water infrastructure in place, such as distribution systems, man-made reservoirs, or natural groundwater basins, determine if and how reclaimed water can be part of the drinking water supply. Some communities reuse water to replenish groundwater basins. Others put it into surface water reservoirs. In these instances the reclaimed water is blended with other water supplies and/or sits in storage for a certain amount of time before it is drawn out and gets treated again at a water treatment or distribution system. In some communities, the reused water is put directly into pipelines that go to a water treatment plant or distribution system.

Modern technologies such as reverse osmosis and ultraviolet disinfection are commonly used when reclaimed water will be mixed with the drinking water supply.

Indirect Potable Reuse

Indirect potable reuse means the water is delivered to you indirectly. After it is purified, the reused water blends with other supplies and/or sits a while in some sort of storage, man-made or natural, before it gets delivered to a pipeline that leads to a water treatment plant or distribution system. That storage could be a groundwater basin or a surface water reservoir.

Some municipalities are using and others are investigating Indirect Potable Reuse (IPR) of reclaimed water. For example, reclaimed water may be pumped into (subsurface recharge) or percolated down to (surface recharge) groundwater aquifers, pumped out, treated again, and finally used as drinking water. This technique may also be referred to as *groundwater recharging*. This includes slow processes of further multiple purification steps via the layers of earth/sand (absorption) and microflora in the soil (biodegradation).

IPR or even unplanned potable use of reclaimed wastewater exists in many countries, where the latter is discharged into groundwater to hold back saline intrusion in coastal aquifers. IPR has generally included some type of environmental buffer, but conditions in certain areas have created an urgent need for more direct alternatives.

Direct potable reuse

Direct potable reuse means the reused water is put directly into pipelines that go to a water treatment plant or distribution system. Direct potable reuse may occur with or without "engineered storage" such as underground or above ground tanks.

In a Direct Potable Reuse (DPR) scheme, water is put directly into pipelines that go to a water treatment plant or distribution system. Direct potable reuse may occur with or without "engineered storage" such as underground or above ground tanks. In other words, DPR is the introduction of reclaimed water derived from urban wastewater after extensive treatment and monitoring to assure that strict water quality requirements are met at all times, directly into a municipal water supply system.

Indirect Potable Reuse (IPR)

IPR occurs through the augmentation of drinking water supplies with urban wastewater treated to a level suitable for IPR followed by an environmental buffer (e.g. rivers, dams, aquifers, etc.) that precedes drinking water treatment. In this case, urban wastewater passes through a series of treatment steps that encompasses membrane filtration and separation processes (e.g. MF, UF and RO), followed by an advanced chemical oxidation process (e.g. UV, UV+H_2O_2, ozone).

Reuse in Space

Wastewater reclamation can be especially important in relation to human spaceflight. In 1998, NASA announced it had built a human waste reclamation bioreactor designed for use in the International Space Station and a manned Mars mission. Human urine and feces are input into one end of the reactor and pure oxygen, pure water, and compost (humanure) are output from the other end. The soil could be used for growing vegetables, and the bioreactor also produces electricity.

Aboard the International Space Station, astronauts have been able to drink recycled urine due to the introduction of the ECLSS system. The system costs $250 million and has been working since May 2009. The system recycles wastewater and urine back into potable water used for drinking, food preparation, and oxygen generation. This cuts back on the need for resupplying the space station so often.

Benefits

Water/wastewater reuse, as an alternative water source, can provide significant economic, social and environmental benefits, which are key motivators for implementing such reuse programmes. Specifically, in agriculture, irrigation with wastewater may contribute to improve production yields, reduce the ecological footprint and promote socioeconomic benefits. These benefits include:

- Increased water availability
- Drinking water substitution - keep drinking water for drinking and reclaimed water for non-drinking use (i.e. industry, cleaning, irrigation, domestic uses, toilet flushing, etc.)
- Reduced over-abstraction of surface and groundwater
- Reduced energy consumption associated with production, treatment, and distribution of water compared to using deep groundwater resources, water importation or desalination
- Reduced nutrient loads to receiving waters (i.e. rivers, canals and other surface water resources)
- Reduced manufacturing costs of using high quality reclaimed water
- Increased agricultural production (i.e. crop yields)
- Reduced application of fertilizers (i.e. conservation of nutrients, reducing the need for artificial fertilizer (e.g. soil nutrition by the nutrients existing in the treated effluents))
- Enhanced environmental protection by restoration of streams, wetlands and ponds
- Increased employment and local economy (e.g. tourism, agriculture).

Design Considerations

Cover for reclaimed water valve, San Francisco Water District

Nonpotable reclaimed water is often distributed with a dual piping network that keeps reclaimed water pipes completely separate from potable water pipes.

In many cities using reclaimed water, it is now in such demand that consumers are only allowed to use it on assigned days. Some cities that previously offered unlimited reclaimed water at a flat rate are now beginning to charge citizens by the amount they use.

Distribution

A lavender-colored pipeline carrying nonpotable water in a dual piping system in Mountain View, California, U.S.

Treatment Processes

For many types of reuse applications wastewater must pass through numerous sewage treatment process steps before it can be used. Steps might include screening, primary settling, biological treatment, tertiary treatment (for example reverse osmosis), disinfection.

There are several technologies used to treat wastewater for reuse. A combination of these technologies can meet strict treatment standards and make sure that the processed water is hygienically safe, meaning free from bacteria and viruses. The following are some of the typical technologies: Ozonation, ultrafiltration, aerobic treatment (membrane bioreactor), forward osmosis, reverse osmosis, advanced oxidation.

Wastewater is generally treated to only secondary level treatment when used for irrigation.

A pump station distributes reclaimed water to users around the city. This may include golf courses, agricultural uses, cooling towers, or in land fills.

Alternative Options

Rather than treating wastewater for reuse purposes, other options can achieve similar effects of freshwater savings:

- Greywater reuse systems - at a household level, treated or untreated greywater may be used to for flush toilets or to water the garden.
- Rainwater harvesting and stormwater recovery- Urban design systems which incorporate rainwater harvesting and reduce runoff are known as Water Sensitive Urban Design (WSUD) in Australia, Low Impact Development (LID) in the United States and Sustainable urban drainage systems (SUDS) in the United Kingdom.
- Seawater desalination - an energy-intensive process where salt and other minerals are removed from seawater to produce potable water for drinking and irrigation, typically through membrane filtration (reverse-osmosis), and steam-distillation.

Costs

The cost of reclaimed water exceeds that of potable water in many regions of the world, where a fresh water supply is plentiful. However, reclaimed water is usually sold to citizens at a cheaper rate to encourage its use. As fresh water supplies become limited from distribution costs, increased population demands, or climate change reducing sources, the cost ratios will evolve also. The evaluation of reclaimed water needs to consider the entire water supply system, as it may bring important value of flexibility into the overall system.

Reclaimed water systems usually require a dual piping network, often with additional storage tanks, which adds to the costs of the system.

Barriers to Implementation

- Full-scale implementation and operation of water reuse schemes still face regulatory, economic, social and institutional challenges.
- Economic viability of water reuse schemes.
- Costs of water quality monitoring and identification of contaminants.
- Full cost recovery from water reuse schemes - lack of financial water pricing systems comparable to already subsidized conventional treatment plants.

Health Aspects

Reclaimed water is considered safe when appropriately used. Reclaimed water planned for use in recharging aquifers or augmenting surface water receives adequate and reliable treatment before mixing with naturally occurring water and undergoing natural restoration processes. Some of this water eventually becomes part of drinking water supplies.

A water quality study published in 2009 compared the water quality differences of reclaimed/recycled water, surface water, and groundwater. Results indicate that reclaimed water, surface water, and groundwater are more similar than dissimilar with regard to constituents. The researchers tested for 244 representative constituents typically found in water. When detected, most constituents were in the parts per billion and parts per trillion range. DEET (a bug repellant), and Caffeine were found in all water types and virtually in all samples. Triclosan (in anti-bacterial soap & toothpaste) was found in all water types, but detected in higher levels (parts per trillion) in reclaimed water than in surface or groundwater. Very few hormones/steroids were detected in samples, and when detected were at very low levels. Haloacetic acids (a disinfection by-product) were found in all types of samples, even groundwater. The largest difference between reclaimed water and the other waters appears to be that reclaimed water has been disinfected and thus has disinfection by-products (due to chlorine use).

A 2005 study titled "Irrigation of Parks, Playgrounds, and Schoolyards with Reclaimed Water" found that there had been no incidences of illness or disease from either microbial pathogens or chemicals, and the risks of using reclaimed water for irrigation are not measurably different from irrigation using potable water.

A 2012 study conducted by the National Research Council in the United States of America found that the risk of exposure to certain microbial and chemical contaminants from drinking reclaimed water does not appear to be any higher than the risk experienced in at least some current drinking water treatment systems, and may be orders of magnitude lower. This report recommends adjustments to the federal regulatory framework that could enhance public health protection for both planned and unplanned (or de facto) reuse and increase public confidence in water reuse.

Many humans associate a feeling of disgust with reclaimed water and 13% of a survey group said they would not even sip it. Nonetheless, the main health risk for potable use of reclaimed water is the potential for pharmaceutical and other household chemicals or their derivatives (Environmental persistent pharmaceutical pollutants) to persist in this water. This would be less of a concern if human excreta was kept out of sewage by using dry toilets or systems that treat blackwater separately from greywater.

To address these concerns about the source water, reclaimed water providers use multi-barrier treatment processes and constant monitoring to ensure that reclaimed water is safe and treated properly for the intended end use.

Environmental Aspects

There is debate about possible health and environmental effects. To address these concerns, A Risk Assessment Study of potential health risks of recycled water and comparisons to conventional Pharmaceuticals and Personal Care Product (PPCP) exposures was conducted by the WateReuse Research Foundation. For each of four scenarios in which people come into contact with recycled water used for irrigation - children on the playground, golfers, and landscape, and agricultural workers - the findings from the study indicate that it could take anywhere from a few years to millions of years of exposure to nonpotable recycled water to reach the same exposure to PPCPs that we get in a single day through routine activities.

Using reclaimed water for non-potable uses saves potable water for drinking, since less potable water will be used for non-potable uses.

It sometimes contains higher levels of nutrients such as nitrogen, phosphorus and oxygen which may somewhat help fertilize garden and agricultural plants when used for irrigation.

The usage of water reclamation decreases the pollution sent to sensitive environments. It can also enhance wetlands, which benefits the wildlife depending on that eco-system. It also helps to stop the chances of drought as recycling of water reduces the use of fresh water supply from underground sources. For instance, The San Jose/Santa Clara Water Pollution Control Plant instituted a water recycling program to protect the San Francisco Bay area's natural salt water marshes.

The main potential risks that are associated with reclaimed wastewater reuse for irrigation purposes, when the treatment is not adequate are the following:

1. contamination of the food chain with microcontaminants, pathogens (i.e. bacteria, viruses, protozoa, helminths), or antibiotic resistance determinants;
2. soil salinization and accumulation of various unknown constituents that might adversely affect agricultural production;
3. distribution of the indigenous soil microbial communities;
4. alteration of the physicochemical and microbiological properties of the soil and contribution to the accumulation of chemical/biological contaminants (e.g. heavy metals, chemicals (i.e. boron, nitrogen, phosphorus, chloride, sodium, pesticides/herbicides), natural chemicals (i.e. hormones), contaminants of emerging concern (CECs) (i.e. pharmaceuticals and their metabolites, personal care products, household chemicals and food additives and their transformation products), etc.) in it and subsequent uptake by plants and crops;
5. excessive growth of algae and vegetation in canals carrying wastewater (i.e. eutrophication);
6. groundwater quality degradation by the various reclaimed water contaminants, migrating and accumulating in the soil and aquifers.

History

Wastewater reuse (planned or unplanned) is an ancient practice, which has been applied since the dawn of human history, and is closely connected to the development of sanitation provision.

U.S.

In the U.S., the Clean Water Act of 1972 mandated elimination of the discharge of untreated waste from municipal and industrial sources to make water safe for fishing and recreation. The US federal government provided billions of dollars in grants for building sewage treatment plants around the country. Modern treatment plants, usually using oxidation and/or chlorination in addition to primary and secondary treatment, were required to meet certain standards.

Los Angeles County's sanitation districts started providing treated wastewater for landscape irrigation in parks and golf courses in 1929. The first reclaimed water facility in California was built at San Francisco's Golden Gate Park in 1932. The Water Replenishment District of Southern California was the first groundwater agency to obtain permitted use of recycled water for groundwater recharge in 1962.

Orange County is located in Southern California, USA, and houses a classic example in indirect potable reuse. A large-scale artificial groundwater recharge scheme exists in the area, providing a much-needed freshwater barrier to intruding seawater. Part of the injected water consists of recycled water, starting as of 1976 with Water Factory 21, which used RO and high lime to clean the water (production capacity of 19,000 m^3 per day). This plant was de-commissioned in 2004 and has since made place for a new project with a higher capacity (265,000 m^3 per day with an ultimate capacity of 492,000 m^3 per day), under the name of Groundwater Replenishment System.

Guidelines and Regulations

International Organisations

- World Health Organization (WHO): "Guidelines for the safe use of wastewater, excreta and greywater" (2006).
- United Nations Environment Programme (UNEP): "Guidelines for municipal wastewater reuse in the Mediterranean region" (2005).
- United Nations Water Decade Programme on Capacity Development (UNW-DPC): Proceedings on the UNWater project "Safe use of wastewater in agriculture" (2013).

European Union

The health and environmental safety conditions under which wastewater may be reused are not specifically regulated at the European Union (EU) level. There are no guidelines or regulations at EU level on water quality for water reuse purposes. In the Water Framework Directive, reuse of water is mentioned as one of the possible measures to achieve the Directive's quality goals, however this remains a relatively vague recommendation rather than a requirement: Part B of Annex VI refers to reuse as one of the "supplementary measures which Member States within each river basin district may choose to adopt as part of the programme of measures required under Article 11(4)".

Besides that, Article 12 of the Urban Wastewater Treatment Directive concerning the reuse of treated wastewater states that "treated wastewater shall be reused whenever appropriate", is not specific enough to promote water reuse and it leaves too much room for interpretation as to what can be considered as an "appropriate" situation to reuse treated wastewater.

Despite the lack of common water reuse criteria at the EU level, several Member States (MS) have issued their own legislative frameworks, regulations, or guidelines for different water reuse applications (e.g. Cyprus, France, Greece, Italy, and Spain).

However, after an evaluation carried out by the European Commission (EC) on the water reuse standards of several member states it was concluded that they differ in their approach. There are important divergences among the different standards regarding the permitted uses, the parameters to be monitored, and the limit values allowed. This lack of harmonization among water reuse standards might create some trade barriers for agricultural goods irrigated with reclaimed water. Once on the common market, the level of safety in the producing member states may be not considered as sufficient by the importing countries. The most representative standards on wastewater reuse from European member states are the following:

- Cyprus: Law 106 (I) 2002 Water and Soil pollution control and associated regulations (KDP 772/2003, KDP 269/2005) (Issuing Institutions: Ministry of Agriculture, Natural resources and Environment, Water Development Department).
- France: Jorf num.0153, 4 July 2014. Order of 2014, related to the use of water from treated urban wastewater for irrigation of crops and green areas (Issuing Institutions: Ministry of Public Health, Ministry of Agriculture, Food and Fisheries, Ministry of Ecology, Energy and Sustainability).
- Greece: CMD No 145116. Measures, limits and procedures for reuse of treated wastewater (Issuing Institutions: Ministry of Environment, Energy and Climate Change).

- Italy: DM 185/2003. Technical measures for reuse of wastewater (Issuing Institutions: Ministry of Environment, Ministry of Agriculture, Ministry of Public Health).
- Portugal: NP 4434 2005. Reuse of reclaimed urban water for irrigation (Issuing Institutions: Portuguese Institute for Quality).
- Spain: RD 1620/2007. The legal framework for the reuse of treated wastewater (Issuing Institutions: Ministry of Environment, Ministry of Agriculture, Food and Fisheries, Ministry of Health).

U.S.

Reclaimed water is not regulated by the Environmental Protection Agency (EPA), but the EPA has developed water reuse guidelines that were most recently updated in 2012. The EPA Guidelines for Water Reuse represents the international standard for best practices in water reuse. The document was developed under a Cooperative Research and Development Agreement between the U.S. Environmental Protection Agency (EPA), the U.S. Agency for International Development (USAID), and the global consultancy CDM Smith. The Guidelines provide a framework for states to develop regulations that incorporate the best practices and address local requirements.

Other Countries

- Canada: "Canadian guidelines for domestic reclaimed water for use in toilet and urinal flushing" (2010).
- China: China National Reclaimed Water Quality Standard; China National Standard GB/T 18920-2002, GB/T 19923-2005, GB/T 18921-2002, GB 20922-2007 and GB/T 19772-2005.
- Israel: Ministry of Health regulation (2005).
- Japan: National Institute for Land and Infrastructure Management: Report of the Microbial Water Quality Project on Treated Sewage and Reclaimed Wastewater (2008).
- Jordan: Jordanian technical base n. 893/2006 Jordan water reuse management Plan (policy).
- Mexico: Mexican Standard NOM-001-ECOL-1996 governing wastewater reuse in Agriculture.
- South Africa Policies: The latest revision of the Water Services Act of 1997 relating to grey-water and treated effluent (Department of Water Affairs and Forestry, 2001).

- Tunisia: Standard for the use of treated wastewater in agriculture (NT 106-109 of 1989) and list of crops that can be irrigated with treated wastewater (Ministry of Agriculture, 1994).
- USA National: United States Environmental Protection Agency (USEPA) "Guidelines for water reuse" (2012).
- Australia National level Guidelines: Government of Australia (the Natural Resource Management Ministerial Council, the Environment Protection and Heritage Council, and the Australian Health Ministers Conference (NRM-MC-EPHC-AHMC)): Guidelines for water recycling: managing health and environmental risks" Phase 1, 2006.

Examples

Australia

When there are droughts in Australia interest in reclaimed effluent options increases. Brisbane has been seen as a leader in this trend, and other cities and towns will review the Western Corridor Recycled Water Project once completed.

While there are currently no full-scale direct potable reuse schemes operating in Australia, the Australian Antarctic Division is investigating the option of installing a potable reuse scheme at its Davis research base in Antarctica. To enhance the quality of the marine discharge from the Davis WWTP, a number of different, proven technologies have been selected to be used in the future, such as ozonation, UV disinfection, chlorine, as well as UF, activated carbon filtration and RO.

Israel

As of 2010, Israel leads the world in the proportion of water it recycles. Israel treats 80% of its sewage (400 billion liters a year), and 100% of the sewage from the Tel Aviv metropolitan area is treated and reused as irrigation water for agriculture and public works. As of today, all the reclaimed sewage water in Israel is used for agricultural and land improvement purposes.

Namibia

An example of direct potable reuse is the case of Windhoek (Namibia, New Goreangab Water Reclamation Plant (NGWRP)), where treated wastewater has been blended with drinking water for more than 40 years. It is based on the multiple treatment barriers concept (i.e. pre-ozonation, enhanced coagulation/dissolved air flotation/rapid sand filtration, and subsequent ozone, biological activated carbon/granular activated carbon, ultrafiltration (UF), chlorination) to reduce associated risks and improve the water quality. The reclaimed wastewater nowadays represent about 14% of the city's drinking water production.

Singapore

In Singapore reclaimed water is called NEWater and is bottled directly from an advanced water purification facility for educational and celebratory purposes. Though most of the reused water is used for high-tech industry in Singapore, a small amount is returned to reservoirs for drinking water.

At the end of 2002, the programme - successfully branded as NEWater - had garnered a 98 per cent acceptance rate, with 82% of respondents indicating that they would drink the reused water directly, another 16% only when mixed with reservoir water. The produced NEWater after stabilization (addition of alkaline chemicals) is in compliance with the WHO requirements and can be piped off to its wide range of applications (e.g. reuse in industry, discharge to a drinking water reservoir). NEWater now makes up around 30% of Singapore's total use, by 2060 Singapore's National Water Agency plans to triple the current NEWater capacity as to meet 50% of Singapore's future water demand.

South Africa

In South Africa, the main driver for wastewater reuse is drought conditions.

For example, in Beaufort West, South Africa's a direct wastewater reclamation plant (WRP) for the production of drinking water was constructed in the end of 2010, as a result of acute water scarcity (production of 2,300 m^3 per day). The process configuration based on multi-barrier concept and includes the following treatment processes: sand filtration, UF, two-stage RO, and permeate disinfected by ultraviolet light (UV).

U.S.

The leaders in use of reclaimed water in the U.S. are Florida and California.

In a January 2012 U.S. National Research Council report, a committee of independent experts found that expanding the reuse of municipal wastewater for irrigation, industrial uses, and drinking water augmentation could significantly increase the United States' total available water resources.

One example is Orange County which is located in Southern California, USA, and houses a classic example in indirect potable reuse. A large-scale artificial groundwater recharge scheme exists in the area, providing a much-needed freshwater barrier to intruding seawater.

Water Treatment

Water treatment is any process that makes water more acceptable for a specific end-use. The end use may be drinking, industrial water supply, irrigation, river flow maintenance, water recreation or many other uses, including being safely returned to the environment. Water treatment removes contaminants and undesirable components, or reduces their concentration so that the water becomes fit for its desired end-use.

Dalecarlia Water Treatment Plant, Washington, D.C.

Treatment for Drinking Water Production

Treatment for drinking water production involves the removal of contaminants from raw water to produce water that is pure enough for human consumption without any short term or long term risk of any adverse health effect. Substances that are removed during the process of drinking water treatment include suspended solids, bacteria, algae, viruses, fungi, and minerals such as iron and manganese.

The processes involved in removing the contaminants include physical processes such as settling and filtration, chemical processes such as disinfection and coagulation and biological processes such as slow sand filtration.

Measures taken to ensure water quality not only relate to the treatment of the water, but to its conveyance and distribution after treatment. It is therefore common practice to keep residual disinfectants in the treated water to kill bacteriological contamination during distribution.

World Health Organization (WHO) guidelines are a general set of standards intended to apply where better local standards are not implemeted. More rigorous standards apply across Europe, the USA and in most other developed countries. followed throughout the world for drinking water quality requirements.

Processes

A combination selected from the following processes is used for municipal drinking water treatment worldwide:

- Pre-chlorination for algae control and arresting biological growth
- Aeration along with pre-chlorination for removal of dissolved iron when present with small amounts relatively of manganese

- Coagulation for flocculation or slow-sand filtration
- Coagulant aids, also known as polyelectrolytes – to improve coagulation and for more robust floc formation
- Sedimentation for solids separation that is removal of suspended solids trapped in the floc
- Filtration to remove particles from water either by passage through a sand bed that can be washed and reused or by passage through a purpose designed filter that may be washable.
- Disinfection for killing bacteria viruses and other pathogens.

Empty aeration tank for iron precipitation.

Tanks with sand filters to remove precipitated iron (not working at the time).

Technologies for potable water and other uses are well developed, and generalized designs are available from which treatment processes can be selected for pilot testing on the specific source water. In addition, a number of private companies provide patented technological solutions for treatment of specific contaminants. Automation of water and waste-water treatment is common in the developed world. Source water quality through the seasons, scale and environmental impact can dictate capital costs and operating costs. End use of the treated water dictates the necessary quality monitoring technologies, and locally available skills typically dictate the level of automation adopted.

Constituent	Unit Processes
Turbidity and particles	Coagulation/ flocculation, sedimentation, granular filtration
Major dissolved inorganics	Softening, aeration, membranes
Minor dissolved inorganics	Membranes
Pathogens	Sedimentation, filtration, disinfection
Major dissolved organics	Membranes, adsorption

Polluted Water Treatment

A sewage treatment plant in northern Portugal.

Wastewater treatment is the process that removes the majority of the contaminants from wastewater or sewage and produces both a liquid effluent suitable for disposal to the natural environment and a sludge. Biological processes can be employed in the treatment of wastewater and these processes may include, for example, aerated lagoons, activated sludge or slow sand filters. To be effective, sewage must be conveyed to a treatment plant by appropriate pipes and infrastructure and the process itself must be subject to regulation and controls. Some wastewaters require different and sometimes specialized treatment methods. At the simplest level, treatment of sewage and most wastewaters is carried out through separation of solids from liquids, usually by sedimentation. By progressively converting dissolved material into solids, usually a biological floc, which is then settled out, an effluent stream of increasing purity is produced.

Industrial Water and Wastewater Treatment

Two of the main processes of industrial water treatment are *boiler water treatment* and *cooling water treatment*. A lack of proper water treatment can lead to the reaction of solids and bacteria within pipe work and boiler housing. Steam boilers can suffer from scale or corrosion when left untreated. Scale deposits can lead to weak and dangerous machinery, while additional fuel is required to heat the same level of water because of the rise in thermal resistance. Poor quality dirty water can become a breeding ground for bacteria such as *Legionella* causing a risk to public health.

With the proper treatment, a significant proportion of industrial on-site wastewater might be reusable. This can save money in three ways: lower charges for lower water consumption, lower charges for the smaller volume of effluent water discharged and lower energy costs due to the recovery of heat in recycled wastewater.

Corrosion in low pressure boilers can be caused by dissolved oxygen, acidity and excessive alkalinity. Water treatment therefore should remove the dissolved oxygen and maintain the boiler water with the appropriate pH and alkalinity levels. Without effective water treatment, a cooling water system can suffer from scale formation, corrosion and fouling and may become a breeding ground for harmful bacteria. This reduces efficiency, shortens plant life and makes operations unreliable and unsafe.

Domestic Water Treatment

Water supplied to domestic properties may be further treated before use, often using an in-line treatment process. Such treatments can include water softening or ion exchange. Many propriety systems also claim to remove residual disinfectants and heavy metal ions.

Desalination

Saline water can be treated to yield fresh water. Two main processes are used, reverse osmosis or distillation. Both methods require high energy inputs and are usually only used where fresh water is difficult to source.

Field Processes

Living away from drinking water supplies often requires some form of portable water treatment process. These can vary in complexity from the simple addition of a disinfectant tablet in a hiker's water bottle through to complex multi-stage processes carried by boat or plane to disaster areas.

Ultra Pure Water Production

Some industries such as the production of silicon wafers, space technology and many high quality metallurgical process require ultrapure water. The production of such water typically involves many stages, and can include reverse osmosis, ion exchange and several distillation stages using solid tin apparatus.

History

Early water treatment methods still used included sand filtration and chlorination. The first documented use of sand filters to purify the water supply dates to 1804, when the owner of a bleachery in Paisley, Scotland, John Gibb, installed an experimental filter, selling his unwanted surplus to the public. This method was refined in the following two decades, and it culminated in the first treated public water supply in the world, installed by the Chelsea Waterworks Company in London in 1829.

Society and Culture

Developing Countries

As of 2006, waterborne diseases are estimated to have caused 1.8 million deaths each year. These deaths are attributable to inadequate public sanitation systems and in these cases, proper sewerage (or other options such as small-scale wastewater treatment) that must be installed.

Appropriate technology options in water treatment include both community-scale and household-scale point-of-use (POU) designs. Such designs may employ solar water disinfection methods, using solar irradiation to inactivate harmful water-borne microorganisms directly, mainly by the UV-A component of the solar spectrum, or indirectly through the presence of an oxide photocatalyst, typically supported TiO_2 in its anatase or rutile phases. Despite progress in SODIS technology, military surplus water treatment units like the ERDLator are still frequently used in developing countries. Newer military style Reverse Osmosis Water Purification Units (ROWPU) are portable, self-contained water treatment plants are becoming more available for public use.

For waterborne disease reduction to last, water treatment programs that research and development groups start in developing countries must be sustainable by the citizens of those countries. This can ensure the efficiency of such programs after the departure of the research team, as monitoring is difficult because of the remoteness of many locations.

Energy Consumption

For many cities, drinking water and wastewater treatment plants are typically the largest energy consumers, having a total of 30-40% of the cities' energy consumption. More than 4% of the nation's electricity goes towards moving and treating water and wastewater. Cost of these energy is consumed in the flocculation basin for drinking water treatment plants and in the aeration basin for wastewater treatment plants. High amount of energy is needed to mix the large volume of water to allow sedimentations to flocculate together. There are current technologies that may aim to reduce this amount of energy. These include optimizing system processes by modifying and improving pumping and aeration equipments.The effectiveness of such technologies are still under discussion as they take up a lot of energy.

Notable Examples

A notable example that combines both wastewater treatment and drinking water treatment is NEWater in Singapore. NEWater is a technology practised in Singapore that converts wastewater to potable water. More specifically, it is treated wastewater (sewage) that has been purified using dual-membrane (via microfiltration and reverse osmosis) and ultraviolet technologies, in addition to conventional water treatment processes. The water is potable and is consumed by humans, but is mostly used by industries requiring high purity water. The total capacity of the plants is about 20 million US gallons per day (75,700 m^3/day). Some 6% of this is used for indirect potable use, equal to about 1% of Singapore's potable water requirement of 380 million US gallons per day (13 m^3/s). The rest is used at wafer fabrication plants and other non-potable applications in industries in Woodlands, Tampines, Pasir Ris, and Ang Mo Kio.

Failures of Water Treatment Plants

When water treatment plants fail, the impact reaches a large group of people. These water treatment plants may fail due to a variety of reasons. They include poor maintenance, power shutdown or the plant may simply not be able to withstand and treat such a high influx of water. This is why engineers often employ safe margin and design for a larger volume than expected. There are a few ways in which citizens can deal with a water treatment plant failure. This includes buying water filtration systems or water filtration tablets.

Regulation by the US Government

Drinking Water

The Safe Drinking Water Act requires the U.S. Environmental Protection Agency (EPA) to set standards for drinking water quality in public water systems (entities that provide water for human consumption to at least 25 people for at least 60 days a year). Enforcement of the standards is mostly carried out by state health agencies. States may set standards that are more stringent than the federal standards.

EPA has set standards for over 90 contaminants organized into six groups: microorganisms, disinfectants, disinfection byproducts, inorganic chemicals, organic chemicals and radionuclides.

EPA also identifies and lists unregulated contaminants which may require regulation. The *Contaminant Candidate List* is published every five years, and EPA is required to decide whether to regulate at least five or more listed contaminants.

Local drinking water utilities may apply for low interest loans, to make facility improvements, through the Drinking Water State Revolving Fund.

Wastewater

EPA and state environmental agencies set wastewater standards under the Clean Water Act. Point sources must obtain surface water discharge permits through the National

Pollutant Discharge Elimination System (NPDES).

EPA sets basic national wastewater standards:

- The "Secondary Treatment Regulation" for municipal sewage treatment plants, and
- Effluent guidelines for categories of industrial facilities.

These standards are incorporated into the permits, which may include additional treatment requirements developed on a case-by-case basis. NPDES permits must be renewed every five years. EPA has authorized 46 state agencies to issue and enforce NPDES permits. EPA regional offices issues permits for the rest of the country.

Financial assistance for improvements to sewage treatment facilities is available to state and local governments through the Clean Water State Revolving Fund, a low interest loan program.

References

- Dickens CWS and Graham PM. 2002. The Southern Africa Scoring System (SASS) version 5 rapid bioassessment for rivers "African Journal of Aquatic Science", 27:1–10
- Finger, Matthias & Jeremy Allouche (2002):Water Privatisation: Transnational corporations and the re-regulation of the global water industry, Spon Press, ISBN 978-0-415-23208-1
- Natural Disasters and Severe Weather. "Water Quality After a Tsunami". Centers for Disease Control and Prevention. Retrieved 2017-04-27
- Hanaor, Dorian A. H.; Sorrell, Charles C. (2014). "Sand Supported Mixed-Phase TiO_2 Photocatalysts for Water Decontamination Applications". Advanced Engineering Materials. 16 (2): 248–254. doi:10.1002/adem.201300259
- George Clarke, Katrina Kosec and Scott Wallsten:Has private participation in water and sewerage improved coverage? Empirical evidence from Latin America, Journal of International Development 21, 327–361 (2009)
- Schiffler, Manuel: Water, Politics and Money. A Reality Check on Privatization, Springer International Publishing, 2015. ISBN 978-3-319-16690-2
- "Center for Coastal Monitoring and Assessment: Mussel Watch Contaminant Monitoring". Ccma.nos.noaa.gov. 2014-01-14. Retrieved 2015-09-04
- International Water Management Institute, Colombo, Sri Lanka (2010). "Helping restore the quality of drinking water after the tsunami." Success Stories. Issue 7. doi:10.5337/2011.0030
- Ran T, Chattopadhyay SK. Economic Evaluation of Community Water Fluoridation. American Journal of Preventive Medicine. January 2016. doi:10.1016/j.amepre.2015.10.014
- Fawell J, Bailey K, Chilton J, Dahi E, Fewtrell L, Magara Y. Fluoride in Drinking-water [PDF]. World Health Organization; 2006. ISBN 92-4-156319-2. Environmental occurrence, geochemistry and exposure. p. 5–27
- "Energy Efficient Reverse Osmosis Desalination Process", p. 343 Table 1, International Journal of Environmental Science and Development, Vol. 3, No. 4, August 2012

- Lobina, Emanuele; Hall, David (June 2003). "Problems with private water concessions: a review of experience". Public Services International Research Unit (PSIRU), University of Greenwich. Retrieved 18 June 2011
- Pizzo G, Piscopo MR, Pizzo I, Giuliana G. Community water fluoridation and caries prevention: a critical review. Clin Oral Investig. 2007;11(3):189–93. doi:10.1007/s00784-007-0111-6. PMID 17333303
- Neemann, Jeff; Hulsey, Robert; Rexing, David; Wert, Eric (2004). "Controlling Bromate Formation During Ozonation with Chlorine and Ammonia". Journal American Water Works Association. 96 (2): 26–29
- Lauer WC. Water Fluoridation Principles and Practices. 5th ed. Vol. M4. American Water Works Association; 2004. (Manual of Water Supply Practices). ISBN 1-58321-311-2. History, theory, and chemicals. p. 1–14
- Steadman, Lis. "East to West: the Future of Water?". WaterWorld. Retrieved 26 January 2012. The figures are quoted from the Pinsent Masons' 12th Annual Water Yearbook
- Horowitz HS. Decision-making for national programs of community fluoride use. Community Dent Oral Epidemiol. 2000;28(5):321–9. doi:10.1034/j.1600-0528.2000.028005321.x. PMID 11014508
- Hatzinger, P. B.; Diebold, J.; Yates, C. A.; Cramer, R. J. (2006-01-01). Gu, Baohua; Coates, John D., eds. Perchlorate. Springer US. pp. 311–341. ISBN 9780387311142. doi:10.1007/0-387-31113-0_14
- Galiani, Sebastian; Gertler, Paul; Schargrodsky, Ernesto (31 August 2002). "Water for Life: The Impact of the Privatization of Water Services on Child Mortality" (pdf). p. 9. Retrieved 10 October 2013

Permissions

All chapters in this book are published with permission under the Creative Commons Attribution Share Alike License or equivalent. Every chapter published in this book has been scrutinized by our experts. Their significance has been extensively debated. The topics covered herein carry significant information for a comprehensive understanding. They may even be implemented as practical applications or may be referred to as a beginning point for further studies.

We would like to thank the editorial team for lending their expertise to make the book truly unique. They have played a crucial role in the development of this book. Without their invaluable contributions this book wouldn't have been possible. They have made vital efforts to compile up to date information on the varied aspects of this subject to make this book a valuable addition to the collection of many professionals and students.

This book was conceptualized with the vision of imparting up-to-date and integrated information in this field. To ensure the same, a matchless editorial board was set up. Every individual on the board went through rigorous rounds of assessment to prove their worth. After which they invested a large part of their time researching and compiling the most relevant data for our readers.

The editorial board has been involved in producing this book since its inception. They have spent rigorous hours researching and exploring the diverse topics which have resulted in the successful publishing of this book. They have passed on their knowledge of decades through this book. To expedite this challenging task, the publisher supported the team at every step. A small team of assistant editors was also appointed to further simplify the editing procedure and attain best results for the readers.

Apart from the editorial board, the designing team has also invested a significant amount of their time in understanding the subject and creating the most relevant covers. They scrutinized every image to scout for the most suitable representation of the subject and create an appropriate cover for the book.

The publishing team has been an ardent support to the editorial, designing and production team. Their endless efforts to recruit the best for this project, has resulted in the accomplishment of this book. They are a veteran in the field of academics and their pool of knowledge is as vast as their experience in printing. Their expertise and guidance has proved useful at every step. Their uncompromising quality standards have made this book an exceptional effort. Their encouragement from time to time has been an inspiration for everyone.

The publisher and the editorial board hope that this book will prove to be a valuable piece of knowledge for students, practitioners and scholars across the globe.

Index